Understanding Humans

by

Daniel A. Shields, MD

authorHOUSE®

AuthorHouse™
1663 Liberty Drive
Bloomington, IN 47403
www.authorhouse.com
Phone: 1 (800) 839-8640

Published by AuthorHouse   08/03/2016

ISBN: 978-1-4208-1055-4 (sc)
ISBN: 978-1-4520-3176-7 (e)

Library of Congress Control Number: 2004098727

Print information available on the last page.

This book is printed on acid-free paper.

Special thanks: Robbo.

Table of Contents

Intro...v

Section 1: Simply, Chaos 1

Fractals..2
Levels Of Organization...5
Forces and Fractals 7
The Pervasiveness of Rhythm...9
Time, and Fractals..10
Time, and Levels Of Organization.................................11
The Microcosm 11
The Butterfly Effect 12
Discoveries, As Chaos 13
Using Chaos 18

Section 2: Behavior, Of Humans 22

The Self...25
Mood..28
Mental Injury 30
Adrenergia...31
Anxiety..33
Panic Disorder...34
Phobia 36
Minor Depression..37
Reactive Conditions 38
Thymic Behavior 41
Major Mood Disturbances 43
Somatoform Disorders 45
Dissociation...49
Personality...52
Schizophrenia..67
Kid Psychiatry...68

Section 3: Functioning, Of Humans.................................73

Identity 73
Substance Use and Abuse 81
Flexion vs. Extension...92

Section 4: Essays, On Humans 103

The Spoils of Capitalism....................................103
The Dummie Syndrome......................................106
The New Enemy: Society Itself..........................108
Don't Feed The Humans.....................................110
Life's Hard, Not Easy 114

Intro

Man, he says, is curious. He reports this boldly to his universe, and in so doing achieves this description by several meanings of the adjective. Perhaps he wonders most why he wonders at all.

He has found the mysteries of his world deep, its recesses frustrating. He plods on, caring not that his tools are crude, nor that he is somewhat blind and deaf. Amazingly, he finds among his discoveries not only that there are fruits to such labor, but that the labor itself is fruit. The map, the treasure. Bit by bit, building on blocks laid over millennia, he has now, only recently, begun to unravel the great enigma of his existence.

Though the sky makes a lot more sense to him now, he's not so sure just how much he knows, and doesn't know. As he gazes afar at these things he can't fathom, he grabs for explanations that he can. He seeks to know his world, and understand it. He needs it, and realizes the converse may not be true. And as his society becomes increasingly complicated, he desperately needs still more knowledge, to guarantee the one thing that matters to the universe regarding the species in the first place, that he survive at all.

Hence this volume. "Understanding Humans" is designed, simply, for anyone who wishes to gain insight into this extraordinary creature. No matter what your level of education, you'll learn something by reading this book. And no prerequisite is needed;

this is the prerequisite, your indispensible guide to understanding humans in The Modern Age.

The author is a primary care physician, doing the job in what they call "the trenches". It is at this end of the profession where, should you so desire, you can get to know, intimately, hundreds of thousands of humans. At "the office". Or in the ER, or in clinics at The U, and the VA. For each and every of them, a story, a unique journey and circumstance. The secrets of their struggles, and the idiosyncracies of their milieu, carry the clues to the workings of us all, and of "all us", together.

"Understanding Humans" is a conversational effort which utilizes that special language called American English. So please, excuse all the neologisms and the occasional bad grammar, and the relative plethora of quotation marks. Though deeply complex issues are presented with pluck and parsimony, be sure of this: "Understanding Humans" is at no times flippant.

The book is divided into four sections. It starts with a very flighty discussion that challenges the reader to begin to understand the world of science's newest "way-out" theory, the one they call "Chaos". It is a new way of looking at literally everything that is and has happened in our universe, so to understand anything, you have to know something about "Chaos Theory". The author admits to having read little on the subject, so what is presented is a fresh view of this fantastically interesting concept. This section, titled "Simply, Chaos", is far and away the book's most important entry in its table of contents. (It may be necessary to surf the internet on a few of the concepts discussed.)

The middle section of this guide gets to the fun of behavior, of humans. Here one can learn to recognize the troublesome actions of relatives and acquaintances, and people in the news, and to understand the roots of their own problems. In a separated section, the functional human is discussed, including insights regarding identity and substance abuse, and a simple narrative on the physicalness of this beast presented as an analysis of the titanic struggle of flexion versus extension.

The back of the book contains a scattering of essays that are meant to inspire thought on an array of modern issues. From "Don't

Feed the Humans", to "Life's Hard, Not Easy", and "The Spoils of Capitalism", and others, there are afforded certain provocations.

They say it's more important to know all the questions than it is to know all the answers. So if you, too, aim to be counted among the curious, then ask questions, good questions, to yourself. Look harder at what you see. Wonder what, and then why. It will require all this and more if among your goals is understanding humans.

Section 1:
Simply, Chaos

It's easy to think that "to look is to see". That to open one's eyes, and take in the universe, be it unsuspecting or otherwise, is simple. That to hear, and listen, and "sense" it, is to have sense about it.

Humans have long recognized that there was something weird about their world, and this impression persists. Elders, around long enough to see and believe, relayed this info to their progeny. They wrote it out on cave walls, and built grand structures to show the world and whomever it may concern that, yes, they too thought there was more there than was meeting their eye.

In school, you can study "science", and practice its "method" to better understand what it really is you are seeing and experiencing. You can learn about the fabulously consequential achievements of some of your ancestors who were particularly attentive observers. Soon, you too will begin to notice that, whatever and wherever it is we are living, it's a real, real complicated thing, and that's for sure.

We are now realizing that the great discoveries of Einstein, Newton, and all the other "Plancks" they stood on, were parts of a paradigm that was, in fact, all of these things together, connected in a vastly complex "multi-level universe". The name of the model is "Chaos Theory".

That term, Chaos, means to say this: that in all systems, no matter how chaotic and random things appear to be, there is always organization to be found. Always. Everywhere. Period. That, according to "Chaos", there is no such thing as chaos. Perhaps you need to look harder, but somehow or another it all makes sense.

Born late in the last century, from the observations and evaluations of a diverse group (sociologists, climatologists, biologists, economists, mathematicians), it is a concept known by another name, as "Complexity Theory", which made the cover of TIME around 1992. Though not technically synonymous, "Complexity Theory" and "Chaos Theory" are talking about this same broad paradigm, or explanation, for what it is we are living and awake to. It is widely considered among the great discoveries of that 100 years of human advancement.

Hence, to be hip to it all, the older term has been preferred by many, such that when waxing philosophical one alludes to it as, simply, "Chaos".

Critical to the development of this model are the mathematicians. It is their formulas and number magic, especially when coupled with the amazing speed of modern computers, that have opened our eyes to how so much of the previously unexplainable can, well, be explained. Alas, we've seen already as students of physics and chemistry that "it always comes down to math" on the most basic levels. Now we're wanting to say the same is true of chaos. Er, Chaos.

This matters to humans. Their instincts as survivor freaks mandate that they learn and know as much as they possibly can. So we start this venture, of trying to understand these "folks", humans, here at the beginning, in Chapter One, by trying to understand what is "going on" in the universe in the first place. This thing called livin', a happening we call "Chaos".

The Fractal

Very basically, it is "the fractal" that is the focus of the Chaos Theory Model. A fractal can be considered to represent any "thing" that has mass and therefore form. That is, every "thing". So, trees and people and everyday objects, as well as "things" like cultures and cities, countries, worlds. Teams. Bands. Armies. In the sea and in the atmosphere, groups of gases and water that collectively act as a "thing" show us again how "things" want to aggregate somehow and form, together, a "thing". We call these units-of-anything "fractals", and the anything they are or become a component of, it's "a fractal" too.

Since all things are so positively unique, we use this term that means "unique geometrical object", or "fractal". That term, fractal, was coined by a German scientist by the name of Mandelbrot, who was making them by plugging mathematical equations into his computer. He showed that if you print this on an X/Y graph, the screen soon would "paint" out some very familiar objects we see in everyday life. He called them "fractals". They are actually sets of numbers, generated by this equation, and are also referred to as Mandelbrot Sets.

So why? How can a mathematical equation be connected to something so abstract as living and non-living objects? Is our existence, and the existence of any fractal, somehow driven and locomoted by a kind of mathematical engine?

In this universe that so baffles us, we should try to look at it and see fractals, and employ all our smarts and insights to explore the deep complexities in even the simplest of "things".

Do The Math

When we studied The Calculus in college my teacher, a lefty, said on the first day of class that people have told him often that the course changed their life in some way. It's true of major math. There's a certain messing with the mind that goes on, an enlightened perspective somehow. It lets you notice it's a pretty nutty universe out there. Or out here, rather.

OK. Let's say I throw my pen at the wall. Now, I can always calculate the thing a certain distance from that wall. It gets half-way there, then half of that, and half of that, and so on, and I can always make that number smaller. So, theoretically, the thing never gets there, but I see it smack the wall, so something's wrong with my theory.

We see that, at the extremes of things, simple laws of math and physics seem to "break down", and no longer apply to the situation. Such "never really get there" asymptotes keep wanting to occur at these extremes. What are they approaching?

Einstein showed it was true of Newtonian Physics. For example, according to the math of relativity, we know we can't speed up a particle too much because, at the speed of light, theoretically, it would weigh an infinite amount. Modern scientists have managed to prove through grand experimentation that the math is for real, that length contracts and time dilates, just as Einstein said. And those particles, barreling down those accelerators, really do get heavier.

We see the oddities of math in numbers like primes, in how they are somehow the building blocks of numbers, even though integers seem so countable as to be accountable. Yet we keep seeing indications that all numbers are not created equal. Can this be?

We've seen a handful of constants that are associated with so many basic things. Pi, Fibonacci's number, Planck's constant, the Ideal Gas Law constant. Bernoulli's constant. What do they mean? Where are they coming from? What about Mandelbrot Sets, and the constant that is in that equation? Are they these numbers?

And consider "1". The "unit circle", with a radius of one, generates the fancy math of trigonometry, based on all that "1" is, and isn't. Waves and circles and cycles, all somewhere in that vast ever and neverland, between zero and "1". When it comes to math, there is no number so significant as "1". (And don't forget that zero, itself, had to be discovered, by the ancient Greeks.)

A fractal is "1" thing, and therefore we must wonder if everything that applies to 1 can somehow apply to the this unit, the fractal. Halved and subdivided in potentially infinite varieties, and yet still "a" single thing.

And what of 2? Pairs, mates, a duplicate. We see things multiplying themselves by themselves, as "squares", a form of 2. A number of great significance, is 2.

Heat energy decreases by the square of the distance from the source, as does magnetism. And the mysterious nuclear force decreases by the cube of the distance, introducing 3. Why is it, the role of such simple numbers?

So, in Chaos, there is much math. It becomes important to know and understand math, because so much we see displays it. Fractals, and the forces that affect them, reek of math, and demand curiosity regarding the connection.

Levels Of Organization

Interaction of objects (i.e., fractals) of similar size identify a scale of activity referred to as a "level of organization". Individual, discrete, similar size-and-form fractals are thusly subdivided at levels from sub-microscopic to galactic scale and beyond. As if to zoom a lens, we can see our transition from level to level, and note the different forces and time-sequences at work in these different levels of organization.

Consider for a moment, the human. As people, we notice we have a certain "personality", evidence of a specific conscious being. We interact within something called a "society", which has its own "personality" and plight, and issues that identify it.

We may choose to hang with a group, a unit contained of individuals with a common enough thread to weave them, however loosely. This unit will take on a "mentality", or identity, as a fractal of folks prone and disposed to whatever, good, bad and everything in between. Simple parties are an example of this.

Occasionally, eventfully, a "gang" of humans will pull an act of "mob mentality", or "mass hysteria", dropping the jaws of observers from without. Together, somehow, they all became "a thing", i.e., a fractal, which acted and "did" what it did. It was created, existed for a time, peaked somewhere, and vanished. Like a cloud, or a storm, or a school or a swarm, or gaggle, units, and therefore fractal. During the time they were together as this unit, stuff happened that wouldn't have had they not so assembled.

We mix societies in a global village, and this same "pattern" becomes apparent. Through war, patriation, and immigration, as well as environmental forces, countries and cultures themselves go through such birth, maturation, and decline. The forces at work at the personal level, the family and friend level, and the local, national, and now "global" levels are all different, and specific to that level of organization of the human fractal.

And the same is happening in the ecosystem that supports them. Plants and animals are being born, maturing, fruiting, and moving on, in the same game as the fancier species that might be chopping or chomping them, as a major force affecting their general fractalling, at this same level of organization. This level, you might say, is the one defined by anything visible to the naked eye and ear, of all that listens and sees.

Let's go to a new level of organization. Earth, itself, is organized somehow into a solar "system", with a center unit that is profoundly large and emits huge amounts of energy, with a number of objects that seem to relate to this "middle". Later we learn that the galaxy is composed of exactly these subunits, solar system fractals you could call them. Like people in a city, or animals in a herd, or air droplets in a cloud, this galaxy, then, a solitary thing, i.e., a fractal.

And apparently even galaxies organize themselves into clusters, and yank and pull on each other as yet another level of organization. We look at gravity as the key force at work in this scale of immense size, and find we know little about most of it.

Now put away the telescope, and get out the microscope. We look inside this "unit", the human, and soon find new levels of organization. We see the individual organs contorted and juxtaposed, and note that while they all are living as one, that they, too, exhibit a certain "personality", showing things they, particularly, like and don't like. For example one may be struck with a certain disease or injury, and may fail before the others.

Microscopically, we see that these organ systems are themselves subdividable as functional units that anatomists refer to as "lobules", where a certain repeating, totipotent group of cells additively perform the function of the whole (fractal) organ. The kidney, for example, has a lobule that is composed of the "nephron", with components such as the glomerulus (the filter), the tubules (for reabsorption of serum contents), and the collecting system, where urine is concentrated. Millions of these make up a kidney. Such lobular basics is true in all our organs.

And lobules, these discrete units, are grouped as "cells", themselves individual units, bound by their cell membrane. Some are round, some are flat, some are real long and skinny. But they are individual, and like the lobule, the organ, and the human itself, they have a personality and a life span that is finite. Some live hours, others for decades. All, fractals within a fractal.

And these cell units, they are composed of "molecules", where a fantastically diverse set of "really small things" inter-play, and combine and separate as "chemical reactions". They are the "cast" of "characters", indeed fractals, at this submicroscopic level of organization.

Molecules, of course, are themselves composed of subunits referred to as atoms, a word that means "cannot be cut". This old term, of course, has proved incorrect, since we can now identify several sub-atomic "things" we call "particles", themselves discrete entities. We know a unique force is at work here. We see it all as yet another level of organization.

So, from subatomic, to molecular, to microscopic, to gross human size, to regional and global and galactic, these are all, separate "playing fields" for existence, and we call them levels of organization.

Forces and Fractals

It should be increasingly obvious that different sets of "rules" or "laws" seem to apply at different levels of organization. Where gravity will toss us hard off a ladder, gravity is a meaningless force at the cellular level, and likewise means little in our interaction within and among societies. And as another example, a tissue's requirements for temperature are much more critical at that level of organization, whereas we can simply put on a coat if we need to.

We know that atoms "react" with each other, and recognize that there is a force they play with called "electronegativity". This force is a measure of the relative attractiveness an atom has for its outer shell of electrons. Weaker elements tend to give up these electrons to elements with a stronger electronegativity, as they combine in a chemical reaction.

At the sub-atomic level, we know that another entirely different "force" is running things, and they call it "nuclear" force. Here, protons and neutrons are magically held together by this very powerful attraction (theoretically, such like-charged entities should repel). And we do know that you can create quite a POP! at our level of organization by cutting some of it loose (nuclear bombs).

Societies recognize forces at work on their individual fractals, the humans themselves, and how the plight of their existence is shaped by them. The role of the family and parenting. Economics. Of media and educational systems. Of proximity in cities. Of birds of a feather, flocking together.

The global villages have still different "forces" at work on them. For example, pro-society forces ("good") have always been noted to be in a chronic titanic struggle with a certain dark force ("evil", or anti-social), and humans have even crafted religious philosophies to guide its members for as far back and as far in the future as anyone can see. Mysterious forces, these. But forces acting on our fractals, for sure.

We wonder what the force is at work in crazy places like solar systems and galaxies, and black holes. Is it simply a gravity thing, or is it a force we can't know how to measure? We wonder, what might make whole galaxies "cluster". Presumably, different forces, for different levels of organization.

The Pervasiveness of Rhythm

Paramount in importance regarding the unfolding of events, and the events that are caused by the forces acting at these levels of organization, is the absolute pervasiveness of rhythm. The simple tendency of all things to begin, accelerate, reach a peak, and then undergo a decline and ultimate demise, or "return" to some baseline, should be considered a huge observation. It is true of nights and days, and light itself is a wave. It's true of sound, and of noise even. It true of wake and sleep. And of seasons, and of epidemics and bug and bird populations. It's true of people, as individuals, and of all things living and having lived. Of stones and mountains and ecosystems, and of cities and countries. And "fads". And worlds and suns and galaxies. We note that, among other things, they have in common this "finite-ness".

Balance sought, perhaps?

We have learned from science of the conservation of matter, and of energy. And of angular momentum (spinning skaters). We know of gas laws and pressure physics, and of social phenomena, and of all things describable by Le Chatelier (dynamic exchange balance in a system). And the bell curve of Gauss. Chaos, in its observations and musings of the fractal experience, notes these cycles, rhythms, dichotomies, and begs to understand this equalizing rule. We see these systems bob and weave and bounce and curl in various manifestations of the inevitabilities of balance, and of "conservation" of everything, and can't help but presume this common thread has this simple implication: balance, sought.

Life, great and small, is a finite experience that essentially comes down to the summation of a whole bunch of chemical reactions burning in some form of balance. We call this chemistry of life "homeostasis", a word that means "balance". Most definitely, it is sought, by the soul and identity that lives any consciousness. It will do what it needs to keep this fire burning, for as long as possible, to stay alive as a fractal unit. It will feed and water itself "instinctively", and likewise fight and claw and scheme and even run like hell, to live another minute.

Time, and Fractals

The role of time in the fractal experience, then, must represent the inevitability of balance and rhythm. During its time on the clock, any and all fractals will live an experience marked by balances and imbalances that will direct the occurrence and logging of events. From its chemical reactions microscopically (managed by the profoundly rhythmic endocrine system), to the jostling for food and nutrients by organ systems, and through the acceleration of youth to prime of life and elderly status, the strings are plucked and will eventually become silent again.

When it's all over, some finite time identification will be tally-able. It will represent uncounted imbalances played out, at all the component levels of organization. Until ultimately, we again can calculate that the wavelength of a lifespan for a fractal is exactly 1, albeit subdivided.

And note that continually and steadily a fractal is evolving and changing, and is never exactly the same thing from one moment to the next. We, for example, are killing off blood cells by the billions and creating new ones, sloughing skin cells and shedding hair, and constantly, subtly, moving on down this "road". The term is "fractalling", yet another brand of tick and tock.

Perhaps these forces somehow utilize time to exert their influence on fractal development. We note they are tied together in the equations of physics. And when we see time show up in the math we can only surmise that events occurring during the life of a fractal are not likely to be random, because time and Chaos are just too complicated and connected for that.

So, a human will go ahead and count seconds, and minutes, and he knows, surely, that he can predict after so many hours, it will be "then". That time is something that "goes on". That tomorrow will be another day, even if his sun blows up.

Time, and Levels Of Organization

Magically, along the way, mankind has made a big discovery about time: it is different for each level of organization. Just like how different forces are at work at different levels of organization, the time elapsed for events to occur differs dramatically. We see this in the amazing speed of modern computers. And in the speed at which gases are evolved to make a bomb. And at the atomic level, we see that it appears to takes "less than a millisecond" for a nuke to go off.

A ribosome will crank out a 100-amino-acid protein in something like 8 seconds, a process consisting of hundreds of individual steps. On the other hand, when you drive out west, in America, a storybook of geology that spans millions and billions of years is played out for you. If it seems like forever, well, that's because you're peeping in on another level of organization.

And then there's the universe itself, with times and distances more impossible to grasp even still.

The Microcosm

There seems to be some math that is trying to tell us that "snapshots" of the fractal experience are offered, as if a node is being reached in its fractal time sequence. An impasse of balancing and imbalancing forces, and a show of hands. It is the microcosm of Chaos.

When Mandelbrot was first churning out these computer paintings of geometrical form he called "fractals", he noted that the thing would progress and eventually reach just such a snapshot, and that the form would then suddenly take off on an all new pattern. What's happening there? Does such a phenomenon occur in the life of such fractals as the living kind? How can you know? Would you recognize it if you saw it?

Has it ever happened to you that, during a game or a relationship or a family outing that some really ironic or otherwise notable something occurred that was perhaps a summation of the nature of things, an inventory of the essence of something or someone or some group's personality or identity? You were floored, maybe, or perhaps more barely noticed at first, but then realized it. Be on the lookout for such phenomena, because almost certainly an important message lies therein.

When people have near-death experiences, they often report they "saw their whole life flash before their eyes". Are they experiencing the microcosm of Chaos? The nature and meaning of such events recalled in this situation are likely to say a great deal about a person. Surely, for all of us just such a "tape" exists, and will someday play.

In their life, and in the examination of their physical structure at any and all points throughout that life, a fractal will display a "story" of the events that went into making it what it is and was, over that identifiable time frame. From inception to end, somewhere in this fractal will lie this documentation, be it rings of a tree, or wrinkles on the face or hands, or in branching patterns, or in the layers of Earth's crust, or in the DNA somehow, this time experience is etched somewhere, and often in several places. Are these displays somehow connected to the manufacturers of fractal essence, with the microcosm itself an event in the fractal time sequence? Can you reach far, and see a connection between the two?

It is possible that, if there is significant imbalance of the fractal experience, the microcosm will be more noticeable to the casual observer. Conversely, more complex phenomena may require an especially keen eye to discern.

Fractal Beginnings
The Butterfly Effect

Chaos people like to talk about a little concept called "The Butterfly Effect" as a model for the idea of fractal beginnings. It goes like this: A butterfly flapping its wings in China sets off a very tiny atmospheric disturbance which then grows and grows, and grows, and eventually there's a hurricane half-a-world away. Preposterous, right? Well, the point still should be well-taken. The fractal starts, maybe, very small. But something has to start it. Then, once begun, what happens during this journey, through the action and timing of forces, will determine if the disturbance set up by the flapping of this tiny insect grows to more than a very local floof.

A fractal has to start somewhere, and that's the point. I remember from my days as a dragster, a large friend of mine explained to me why people get their piston connecting rods "polished" for high performance motors. He dramatically pointed out that the rods wouldn't be as likely to fracture because "there wouldn't be no place for a crack to start!"

So, importantly, the following question: Does every fractal have to have something like a butterfly to start it? In Chaos, the very beginning is so consequential that we must ponder what all goes into that. For humans, courtship, and romance, seem so natural, but in fact must involve the co-incidence of so very many things over a very long time.

Why are "mates" so often the nature of things in the fractal universe? Presumably, it is this concept of balance, and that nothing can occur or exist unless it is somehow in a balancing act. And not just in humans, but in virtually all animals and plants too. In society itself, as positive (social) and negative (personal/instinctive) behavior. In weather, as highs and lows. In the sea and on land, with prey and predator. With plant (make) and animal (take). And even in the cell, a double helix. Fractal mates, themselves microcosms of the "goo" of life, managing the chemical reactions we "live".

Discoveries, As Chaos

Occasionally, humanity generates an especially super duper mind. One who might, for example, become obsessed by the konk of an apple, or fascinated by the fantasy of riding on a light wave. People who look, and see, and open the door for others to be so enlightened. We name laws of nature, and elements, and schools after them. We learn, and live their discoveries.

Usually, they are finding ways to measure, and therefore predict, how forces are affecting fractalling at different levels of organization. Ways to use constants, and primes, and 1 and 2. Integers. Squared numbers, and cubes. We think of Mandlebrot, and fractals. And Chaos.

Exhibit A: MODERN PHYSICS Newton became famous for the discoveries of math in everyday things, like objects in motion, battling gravity and inertia. There's momentum, power, work. These are the maths of things that are happening at our degree of bigness. When refuted by Einstein, humanity became aware that different forces and time sequences were applying to different levels of organization.

The proposals of Niels Bohr, Linus Pauling, and others, were the forces of electronegativity (chemistry) and nuclear force (quantum mechanics), the happenings of the very small. When we looked outward instead, from Palomar and elsewhere, we gazed at yet another level of organization. Gradually we became struck by the similarities.

Exhibit B: GAS AND LIQUID If you study gas and fluid physics, sure enough some math emerges. Constants. Squares. Things are moving, so there is distance and time to consider. This must be Chaos. Can we see levels of organization?

Answer: yes. Fluids and gases, when left to fractal observation, will show organization. Water in a pipe will organize into sheets and layers. And in a pond, the water will find a way to organize itself into masses of various temperature. If enough water and depth is there, the winter will cause the cold top water to plop to the bottom of the pool area, as if a fractal on a mission.

In the oceans, we see currents and El Nino and the like, as well as a constant cycling between evaporated water, precipitation as liquid or solid, and variable storage on the land. In Earth's atmosphere, rather amorphous globs of gas will show fractal qualities. Almost amazingly "systems" will seem to "form", get fanned by some butterfly mitigation, and soon "experts" will be trying to predict when they will peak, and begin their ultimate demise.

The dichotomy in weather is played out, often spectacularly, as regions called "highs" and "lows". In the former, the air is dense, rotating clockwise, and moving toward the Earth, while in the latter the air is relatively light, rotates counterclockwise, and is rising away from the planet. Both will form, grow, reach a peak, and then unravel and end after some time frame. And when they interact, they manufacture smaller "systems" that will similarly start small, grow to some form of maturity, and die out as well. Fractals, within fractals.

So we stand there, on our back porches, and watch the weather, noting patterns of organization. Sheets of rain. Squalls. Tornadic activity. Cloudage. And if we walk in and turn on The Weather Channel, they will show you what it looks like from space, how the whole big picture is unfolding. A peek, of course, at fractal weather phenomena, from afar, on a global scale of organization.

Did you just feel a drop? The molecules, themselves fractals, are now grouped to form these larger fractals, some as liquid, real small like that. The drops are forming by starting as a small cluster of molecules, only to be under forces that favor the growth of this fractal beginning, and then eventually it exceeds its point of precipitation, and is therefore dragged to the surface by Earth's gravity. Only when the "unit" was big enough did the planet want it back. (Actually, this occurs when the force of gravity exceeds the lofty updraft of the storm.)

They're just this tiny part of something as big as a region, or the size of a state, or a county, or even a campground. And, regardless, fractal entities. Discrete, individual. All of them.

Exhibit C: THE TREE FRACTAL Conspicuously ubiquitous, the form of a tree. See the math? Is there something to simple bifurcation? Whatever it is, a lot of things organize into this simple concept. From oaks and maples, to really most plants in some stretch of the imagination, along with drains like the rivers of a continent, and blood vessels and kidneys and lungs in an animal. There are family "trees", and indeed the history of life on Earth can be illustrated as a large "tree". Why that form?

If you put in the keyword "Fibonacci" on your searcher you can read on any number of sites how there is so much identifiable math everywhere around you. And this great scientist observed simply this: a sequence where you add two numbers to make a third, then add that number to the one before it in the sequence, and so on. It begins as 1, and 1+1 is 2, so 1, 2. 2+1=3, making the sequence 1, 1, 2, 3. 3+2=5, giving you 1, 2, 3, 5. Then, 5+3=8. The sequence becomes: 1, 1, 2, 3, 5, 8, 13, 21, 34, 55, 89, 144, and so on. These are called "Fibonacci" numbers. The geometric shape they form is a spiral.

And they're literally everywhere. It's a key bit of math, this sequence, at our level of organization, and presumably is intimately tied to some force that is driving the fractalling on our "scale". Sprout, grow, flower at the advancing end. The "new" and "fresh" end. Note the difference between the base, or trunk, of the tree, and what the structure becomes out at its end, at the leaves. While the trunk, including the roots, offer water and ion supply ("inorganics"), the leaves exchange gas and manufacture the complex molecules of life ("organics").

Figuratively, we recognize this tree form because of the obvious functional dichotomy presented. Leafs, and fruits, and "roots" of matters. It should not be surprising, this relationship. They are connected by more than limbs. And don't forget, all things tree must start as a seed, and fetch a favorable Butterfly Effect, while not ending up in the gut of some rodent.

The nervous system of creatures is itself a tree fractal. There is the "trunk", the spinal cord, where a rather simple neuroanatomy carries sensations and stimuli for extremity movement. Just inside the head, there is the "brainstem", which has basic survival electricity for breathing and blood pressure, the rhythms of life. It is here that consciousness is generated.

Just after the brainstem is the midbrain, where hormonal patterns are managed, bringing us homeostasis and therefore sustainable existence. On the way to the forehead are centers for limb feeling and movement, to speech and hearing and more primitive behaviors, until eventually arriving at the personality and memory centers of the heart of you, the brain's frontal lobe.

This map, a tree fractal, from tail to forehead, is no less than the document of animal evolution on Planet Earth. This snapshot is the physical relative of the microcosm of Chaos mentioned earlier.

The human, most obviously through puberty, is a tree fractal that grows from the bottom up. First the feet get big, then the legs and pelvis. Family members are noticing this "growth spurt". Eventually, the arms grow and the shoulders broaden, at full growth around 20.

The brain is also maturing through this period, and likewise does so from the bottom (the back of the head) to the top (the front of the head). In early and middle teenhood, the drama of hormonal patterns will be particularly noticeable. Next, physical movement and coordination will impress us with their athleticism. And poor judgment, thrill seeking, with this hormonal enhancement, may mystify parents and teachers, who know that, generally, once the frontal lobe, behind the forehead, finally develops, a more "reasonable" and "decent" human, with more character and insight, will be represented.

Exhibit Next: THE BALL FRACTAL If you splash water in a pool, and look at the droplets in the air, they're spherical, essentially. Students of physical chemistry know that this is because a sphere is the minimum surface area form for any given volume of stuff. Since liquids can change their form so simply, it's easy for them to morph into this general shape, a comfortable shape for them.

We know about the math of circles and spheres, and that constant called "pi". The area is a squared number. The volume of the thing in 3D is a cubed number. This is beginning to make us think of Chaos.

We know of the useful equations surrounding "the unit circle", and trigonometry. We note the tendency of a "center" of round and spherical things to be so popular. The nucleus of atoms. The nucleus of cells.

Baseball players know when they hit a ball on the "sweet" spot of the bat, which is the thing's center of gravity. Pow. Gone. It's as if the whole of the bat's weight was directly lined up with the whole of the ball's weight. Center on center. It's why we watch.

Dictators. Ringleaders. The center of planets and suns. We see galaxies wanting to be versions of round and circular, and rotating. So often, a "center" of things. Why do things want to have a center? Is God at the center of goodness, and some "Satan" at the center of badness? From queen bees to alpha males, to the eye of the storm, the nidus of infection, to the "centeral" nervous system, and even black holes, why such "centering"? Is it a remnant, or document, of the butterfly effect of fractal beginnings?

Agreed, these are obtuse and diffuse gleanings of but a few of the many questions posed by a fantastically complex universe, and challenge you to look, and perhaps only begin to see, the Chaos in chaos.

Using Chaos

This may seem like a bunch of "way-out" stuff to some people, but in our particular peer group it is part of the commonspeak. My brother's bowling team is called "Mandelbrot", and they won the league last year. He said that, for some reason, everybody bowled bad against them, and I personally had my worst night ever when we were on the same pair of lanes. So I named this year's team "The Fractal", and, well the plack's down on the mantle, and there are microcosmic goodies that happened to come with it. Such fun.

I was one of the last to see the video "The Colors of Infinity", a PBS show about Julia, Mandelbrot, and fractals, and levels of organization. The tape ends with a guy saying, "You often hear, of a treasure buried somewhere, and a map that leads you there. Well, in this case, the map is the treasure."

The computer is also there for your assistance. All the cool and hip kids have heard about Chaos. Fortunately for me, I've had the help of a few really tuned-in lefties. They look, I see.

Back in the '70s the U.S. military created the "fractal analyzer" which allowed it to gaze "into" photographs from space during the Cold War. Fuzzy dots soon became bombers and missiles, and high-level espionage went to another level. This is perhaps the earliest application of fractal form and development in the modern era of understanding.

But in a more basic way, the implications and utilizations of Chaos are deep and far-reaching. Not only can Chaos tell us why and where we've been, but also it is of great value in following, evaluating, and managing phenomena once they have clearly "started", and invites us and challenges us to apply appropriate forces with appropriate timing to favor certain aspects of fractal development and, importantly, non-development.

We must recognize the importance of events that occur early in the fractal experience. The simplest example is in personality development and interpersonal behavior. We notice how events that occur early in life set up the attitudes, insights, and beliefs a person will employ in his behavior in the society. And, as health care professionals, we see the magnitude of the mental suffering and physical symptomatology as these victims of nurturing failure plod their way through a life not worth living. Interventions early in fractal development of a positive nature allow for a creature much more adept at coping skills, adaptation, and goal-directed adult behavior. If we desire a better society, we put all our smarts and resources into appropriate shaping of our young fractals, the earlier the better.

We know that with many if not most diseases and ailments that early intervention is critical in keeping processes from getting "too far out the fractal", becoming so big as to be an unstoppable problem. We adore "prevention", of bad fractals of course. We nip things in the bud, and pull "weeds" when they're young. We urge ounces of prevention, versus pounds of cure.

We watch as epidemics frighten us with their arrival and spread, and rest with at least the assurance that all of them will go the fractal way: acceleration, peak, and eventual demise. We do what we can to make the peak happen as soon as possible.

When people are sick, we apply certain forces (fluids, antibiotics, anti-virals, anti-inflammatories) to blunt the burgeon of balance-threatening conditions, and figure that usually we can get them over the "hump" of a bad fractal so the person can, with the upper hand, get back to the balance of wellness.

We use Chaos already in our attempts to forecast weather, by predicting how systems tend to fractal on, and out, based on their size, temperature, and pressure. And, of course, it is inherently inexact in prediction, and much clearer in retrospect. Locally, the inability of a human to recognize a dangerous fractal escalation or conclusion can result in death by lightning. Each year, more people are killed by lightning than by all the tornadoes and hurricanes combined.

If you're "smart", you can use Chaos at the casino. Ride the good fractals until they start to fade, and then go and sniff out another one that's growing and accelerating. Take home their millions.

Watching economies is quite the chaotic thing. In ours, for example, we only know that it must continue to grow, like any fractal. If it involutes even a little, all hell seems to break loose. An entirely different midset takes control. A different fractal, growing in a "bad" sort of way. Consumer "confidence" is said to have been lost. So we apply certain forces to redirect the thing, before it goes "bad".

In geology, we look back at a fractal level of organization that has gone on for billions of years. We look at plate tectonics, and continent development, and see fractals. We know they fractal on today, offering a constant threat of tremblers at famous hot spots on the globe. We sweat the realities of the time frame of another level of organization.

We've seen how ice-ages have come and gone, with glaciation and it's scaping of the land. With periods of melt and flood and freeze and snowcappery. For sure, fractal levels of organization. The main force? Light, from the sun, yet another fractal, doing some wild thing on a level of organization that is way, way beyond us.

And, alas, we see it in the human condition. Whole societies, that grow and grow, and peak and ultimately fade. And the creature itself. Can you see its "evolution", its gradual, fractalloid differentiation? We don't go to gladiator death matches anymore, as whole societies. And we're getting bigger as a creature, each of us several inches bigger than the average human even two or three hundred years ago. Athletic records, sure enough, continue to fall. It would be a bad sign if they stopped.

In America, the creature is getting much bigger, and also fatter, lazier, dumber, and more self-minded, on average, over the past several generations. Can we see evidence of some form of that most magical of forces, acceleration? The Modern Age itself hath brought on so many new forces (food supply, informationism, lawyerism) that the way most of the creatures learn to think shows questionable logic. One thing's for sure: beware the prognosticators speculating on levels of orgnization of the human fractal.

Looking back through the eyes of the modern Chaos scientist can be fun. All those things they taught in college in the '70s, when I was in school, look so different. Miracles of meiosis as random events?

Doubt it. The pessimistic pervasiveness of entropy? Not seein' it. Population predictions? Please. Evolutional biology? Never got the whole story. At least we saw math everywhere.

Regarding history, could the Nazi's have been stopped at Alsace-Lorraine? Did Kennedy dodge a nuclear exchange by calling out the Soviets about the missiles of October? Was Viet Nam a big gigantic waste of life for absolutely nothing, just because a fool or two couldn't recognize a bad fractal, and properly predict which direction the dominoes would actually fall? Bomb Saddam to avert a more catastrophic Armageddon? Did two superpowers yoke and create, unwittingly, Osama?

Is there power in positive thinking? Do you get luckier the harder you work? Is your pillow more comfy after you've busted your butt a little bit? Is this why it appears that life is hard and not easy?

The Lord works in mysterious ways, but the devil is not at all subtle. Are these and other "spiritual" observations evidence that humans through the millennia have marveled at such complexity? From chariots of the gods, to cathedrals for a God Almighty, to Karma, Manna, Zen, and million other isms long extinct, they all had the same thing in common: order in chaos. Or, simply, Chaos.

Section 2:
Behavior, Of Humans

Do you notice people's behavior? When you're annoyed or floored by the things they do and don't do, is your reaction dismay and disappointment, or observation and evaluation? Anger and angst, or amazement and amusement? Do you try to fix them, or avoid them?

In its big brain a human "creates" and "imagines" through memory and learning, reaching conclusions and prognostications by what is referred to as "logic". The organ must also process input it is receiving from the body itself, and the environment it resides in and subsists on. To "feel" and sense, to follow impulse, to do.

It aims to expel thought through action and dialogue, and move on to the new and important. It prefers to feel what it wants to feel, and will seek such things. If its logic ever seems ill, a closer look reveals it doing just exactly as it wishes, means to ends.

The adult mind will reflect the "programming" of the years it was being educated by its most unique and singular experience during that crucial period at the front of life we call "the nurturing process". Through its use of the tools "learned" during this time, the "grown" human will react and respond to the events and people it encounters.

We note a broad spectrum in relative complexity of behavior among our population. We wonder why some people seem to be able to use more of their gray matter than others. We find it natural to loathe thought poverty, as a door not open, or not opened. Or slammed shut. A brain consequently satisfied to under-"achieve", and "underperform", giving us all witness to this variance in social capacities.

The result, is all those mall goons and reunion loonies, and weirdos in the workplace. Which is to say, all of us. Chapter One of the psychiatry book points out an important observation: the prevalence of psychopathology in the general population is drastically underestimated. From family and friends to colleagues and coworkers, identifiable misbehavior is frequently on display. It affects the quality of life for all of us, and especially for those people who have to live and co-exist with people who can't act right. And it brings to the media a constant supply of drama and melodrama.

As Modern Age humans are more and more exposed to, indeed bombarded by, the diversity of opinions on literally every aspect of life, the ideas created by this varied input is resulting more and more in overtly bizarre and illogical behavior. Like so many hysterical monkeys, on we go, upward and outward. Way outward.

With society posing an ever-more intimidating, frightening, and overwhelming entity, the human of today spends more and more of its time "integrated" into this maelstrom, just…consuming and being consumed. In fact, he and she need to know quite little to get by, to survive in this tidal machine of civilization. Coupled with the casualty of affluence, the bleakness of disenfranchisement, the discouragement of divorce and parental and clan non-support, there results this lot of lessons for learning not demonstrated, and mis-direction never recognized.

Indeed, that which "goes on" behind the closed doors of the great American Everywhere we may not know, but when we see the "product", we'll wonder.

Thus, while it helps to have a decent IQ and a reasonably balanced mood, it is apparent that we all must "learn" how to act. We watch those around us, who impress us with something they do, and we copy them. We pluck little features of those people and create a "personality", displaying the beliefs and attitudes that we identify with. "Apes", that's us, definitely.

Most important of all in the psychological and social development of everyone is the profound importance of the two who yoke to create a human in the first place: the parents. To the developing child they serve as the one sure antidote for the poisons offered by the increasingly invasive forces of their world. They all do right things and wrong things, but more importantly they must simply be there, in the lives of their offspring, living and growing with them. A difficult task, for sure, but for the most part society pays for the sins of dear old mom and pop.

Talking Psychiatrically

With all that in mind, let's talk psychiatrically. From here until the section on "adrenergia" some of the more basic lingo of behavior are highlighted. Perhaps such jargon will make you more hip to the quirks of this nifty, crafty creature.

Firstly, for the most part, behavior is rhythmic like everything else. We all recognize good periods and other times when we don't seem to do as well, or be as sharp. We feel and live these roller-coaster rides of peculiar and particular capabilities, and may not notice or pay much attention to them.

Sometimes, the going gets uniquely difficult. While most of us weather these dark times of our existence, others decompensate instead, where for them a prolonged period of some form of mental struggling unfolds. When such a pattern of abnormal behavior is sustained for weeks or months, we say you are having "an episode". From the dramatic episodes of bipolar affective disorder (mania/depression) and schizophrenia, to the more subtle anxiety and depressive ones that at least half the population will experience, these will usually run a course and return to a more baseline performance, even if nothing is done. Some episodes go several months, few longer than two years.

"Insight" is the term used for the assessment as to whether the patient agrees that his or her behavior is abnormal. Some people have none, while others are terribly bothered by the way they're thinking.

"Vital sense" is how a person thinks they feel physically, and emotionally as well. It may range from energetic and can-do, to tired, weak, and hurting. "Self attitude" is a person's gauge regarding how they think they are perceived by others, and it ranges from competent and groovy, to luckless and down and out.

A few of Dr. Freud's terms are helpful when talking psychiatrically. His concept of "ego" seems to mean this: there is, not surprisingly, a dichotomous relationship in "the mind", with primal, self-protective thoughts and instincts, even to the point of exploitation, on the one side ("the id"), while on the other a more progressive, socially altruistic set of thoughts and instincts which he called the "super-ego". Presumably these two entities are in some sort of dynamic battle as mates in the mind, and it is left to what is called the "ego" to present to the world the actual behavior we all see.

In more "normal" individuals the ego shows logical and objective adult behavior, balances primal and social instincts, and learns and stores memory appropriately. In others, something drives this "ego" to become "overactive", and instead of normal behavior we see a created, deceptive, "phony" performance. Such maladaptive schemes are being utilized by the mind presumably as a diversion from embarrassment or extreme social pressure.

We say the ego is trying to protect the mind, and we refer to these schemes as "ego defense mechanisms". Despite their blatant nature, the person suffering the behavior is unaware of the negative impact of what they are doing, and the term "ego-syntonic" is used (i. e., there is no "insight"). It is a behavior that is "comfortable to the ego".

The Self

A human is a single, discrete entity that has things it likes and doesn't like, wants and doesn't want. It is this quest, instinctive and natural and primal, that is the essence of life: to provide for the self that what it so desires and requires. And for sure, it is quick to defend.

As we grow and mature into modern adulthood we develop skills to use when interacting with each other. These are more sophisticated, learned behaviors. When we perfect them we invite into our lives positive energy and concrete goods that are the bounty of effective, motivated, cooperative socialness.

Unfortunately, a lot of people never figure that out. They are for whatever reason unable to overcome the instinctive sense of self, of perceived immediate needs of this self, and will resort to dastardly maneuvers to serve this self. In so doing, generally what is accomplished is the ultimate defeat of this self.

It is thus a very common and simple thread in so many flavors of errored behavior: too much focus on the self. In varying degrees all of us employ evasive maneuvering when we stumble upon uncomfortable circumstances in our social realm. We call them "ego defense mechanisms", and perhaps one of the world's wrong-acting people has pulled one of them on you. Freud would say "the ego", attempting to mediate between the id and superego, is "utilizing" one of these schemes of behavior to, best it can, further the cause of the self.

Seen these ever?

*Denial—This easiest of first impressions is something we all must overcome. Some never manage, to absurd degrees.

*Rationalization—Often, dealing with failure engenders the desire to blame another person or mitigating circumstance. The worst of lost minds can rationalize essentially any self-serving behavior they like.

*Projection — Accusing someone else of something you're guilty of, such as something a person has done that is embarrassing or shameful. A husband accuses a wife of cheating, when he's the one who's been unfaithful. Someone who is incompetent accuses someone else of it, and so on.

*Intellectualization—Creating a diversion by over-interpretation or use of twisted philosophy. A "smokescreen". Over-talking and over-mooding.

*Compensation—It may seem natural to cover weaknesses by emphasizing the upside of the self. But if there is consistent refusal to identify shortcomings of the self there may result a snowball of personal litter never managed, with bad behavior the consequence.

*Sublimation—From wife-beater preachers to muscleman cops, and bouncers, to the vicarious thrill-seeking "rescuers", the "take" is clear and obvious. (All of these mechanisms are clear and obvious.)

*Turning against the self— Self mutilation, home-made tattooing, and suicidal gesturing are surprisingly common coping mechanisms for people who carry emotional baggage from disasters suffered during the nurturing process. The anguish resulting from terrible memories causes so much distress that one can actually feel better by inflicting this hurt, even if it's on themselves. Some dissociate (enter a new identity) when they "cut" themselves.

*Displacement— As opposed to self-mutilation, some people, you may read, take their pain and un-holdable aggressive instincts out on their fellow man. Especially when directed toward a stranger, such "acting-out" is a way for the id to loose some of its dark energy, energy manufactured and driven by the same old fuel, abuse and neglect, and failure of nurture.

*Somatoform Behavior— Performances like hypochondriasis and others are examples of where imaginary illness becomes a means to an end as simple as wanting to be cared for or cared about.

*Dissociation— An apparent disorder of memory integration and identity, a mind may appear to have no knowledge (no memory) of an event or events that were too painful to bear, and remembering them profoundly undesirable.

If you can recognize these performances when they are perpetrated, you can safely conclude that somehow, somewhere, a disaster of nurture has occurred in this person's upbringing, and hurt the construction of their personality.

Freud also indicated that there were key stages in the developmental process, and that if one of them were not reached, or was interfered with somehow, that bizarre behavior would result. From separation issues of infancy and early toddlerhood, to potty training and control of motor skills, to development through identity and sexuality, these are the doors opening, and in need of opening, in order to get to new levels of personal and social behavior. The fixation that results from such arrest of development results in this abnormal performance, and perhaps inhibits in some way the ability of the conscious mind to utilize its many centers of gray matter. Hence, odd behaviors are simplistic and stubborn.

Particularly, the development of sexuality and sexual identity appear to be of critical importance in the overall maturing of the mind. Premature introduction to these issues, particularly in any kind of physical way, can drastically derail one's ability to turn out normal. It is obviously a very big stage or level that is, hopefully, reached and passed through on the way to finding the self.

Mood

Ever notice that some people just seem to have more pep and pizzazz than everybody else? Or, less maybe? We call this basic phenomenon of behavior "mood".

Just inside the base of the skull the spinal cord begins to widen like the stalk of a cauliflower head. It is this area, the brainstem, that provides for the brain itself its most basic of necessities: an "on" switch. Like a rheostat on a dimmer, it "lights up" the gray matter above, and we become "awake". Here, next to the adrenaline center, is the generator of conscious thought. The relative energy level that it provides is what is called the "mood".

In high moods the person has a lot of energy, talks a lot, feels good physically (has a high "vital sense"), and approves of his behavior (a positive "self-attitude"). Ideas are plentiful, and there are "plans", and goals that are felt to be reachable. Positive thoughts are "sponsoring" positive thought.

Importantly, in high mood there may be "good" and decent behavior, or there may be irritability and dramatics. Either way, a more assertive "make things happen" approach to matters will be evidenced in activities of daily living during high-mood states.

During periods of low mood a person will express negative thought, and hopelessness, and complain physically, particularly of fatigue and pain. There is passiveness, and dependency, and a loss of interest in pleasureful activities. Appetite and energy levels are poor. Restorative sleep is difficult to come by. These moods bring a gloomy and pessimistic attitude, or what is called the "poor outlook", of "poorly sponsored thought".

Since everything in the known universe has a rhythm and balance to it, the same is true of the brain's mood performance. Its amplitude and frequency vary from human to human. The best guess here is that for most people the mood wavelength is about a month or three. But surely, for all of us, it's up and down we go.

We use the term "depression" when you feel bad, think bad, and are sick a lot, or in other words, are in a low mood state. We estimate 40% of office patients and as high as 70% of inpatients have a measurable amount of such depression. We have long seen this connection, and assume that disease is some combination of pathology (germs, inflammation, malignancy) and mood shortcomings. Good nurture, and its great gift, a high vital sense and self attitude, is the preventive Rx for disease of all kind.

"Delusion" is when a person sees one thing and thinks another thing happened, an illogical interpretation of events. It occurs at the extremes of mood. In "mania", the classic high mood state, such delusion might include omnipotence and omniscience (grandeur), or paranoia and persecution. The brain seems to be over-driven by a very active, dominant set of thoughts and emotions, fueled by this big volume knob we call "the mood".

Auditory hallucinations (brain-manufactured voices, often giving instruction), ideas of reference (the people in the football game huddle are talking about me), and rarely, visual and tactile (feel) hallucinations (brain-manufactured sensation) are more examples of an abnormally high mood condition, where thoughts and sensations are literally created from nothing by this brain so over-stimulated by its mood drive. Often sustained for months, it is the classic "manic episode".

Extremely low moods are fraught with psycho-motor retardation (vegetative states), illness behavior (dependency), and sick-role playing. And very commonly, suicidal ideation and gesturing (rejection of the self). In the fascinating condition of "major" depression, such deep reduction of the mood level occurs that near complete vegetation is seen, to the point of not bathing, not eating, and expression of very little thought.

"Psychosis" should be a word that means a chemical neurotransmitter pathway is out-of-whack somehow (nature?), and the result is disorganized and pointless thought. We argue that abnormally high moods (mania) and low moods (psychotic depression) are conditions caused by such "chemical imbalance".

Another example of such a biological error is the classic psychosis schizophrenia, a disorder marked by months-long episodes of delusion and hallucination, to the point of eventual incapacitation. The psychotic break they suffer displays disruption of thought on all its levels: delusion (content of thought), hallucination (perception), inappropriateness of affect (social thought), logic problems and loosening of association (form of thought), preoccupation and fantasy (autism), inability to establish goals (ambivalence), and abnormal posturing (physical thought).

These basic features of schizophrenia display the secular nature of the mind. The eight or so building blocks of behavior need to be appropriately and evenly stimulated from the generator, these "lower centers" down in the brainstem. They need to work together and show balance, because when they don't, we act funny.

Mental Injury

More often than not, minor misbehavior and misfiture are the result of some sort of failure of the nurturing process. Verbal assault and undermining tend to lead to compulsive and controlling behavior, and passive aggression. Physical abuse leads to violent and hateful individuals. Exposure to parental anger may lead to offspring who want to exaggerate themselves. Sexual abuse consistently fosters hate of the self, a sense of chronic pain, and rejection of the identity. Exposure to violence against the mother can generate some of society's most dangerous people.

These are examples of mental injury. The brain, now so maligned, hurt so early, can never be expected to find its way, leading to maladaptive behaviors that are life-long. Uncovering "abuse histories" represents a very important step in the treatment and management of those so-injured.

"Neurosis", an old word in psychiatry, was meant to indicate that a behavioral disorder has resulted from mental injury. It is treated through thought re-manipulation, or "therapy". This type of work may take years and still not "cure" a person, but is more of an on-going assistance to a brain that is unable to ventilate and get past thoughts it is generating, thought generated by deep and dark memories, from long ago.

In minor ways, a lot of us can trace a quirk or an avoidance we have to something unfortunate from our past. There are a zillion moments that shape us, and some lousy ones in there for sure. But for many, a particularly horrendous experience, usually at the hands of a "loved" one, left them broken and limping, mentally, through their existence.

It is common for the more severely maladaptive individuals of the world to express that they have some sort of "chemical imbalance", as if to absolve themselves from self-defeating behavior. Medications, however, will typically have had little effect on their plight, whereas with manics and schizophrenics, conditions where there really is a chemical out of level, they tend to respond well to medications.

Hence, much of behavior is this complex mix of nature and nurture, and separation of the two's effect on the final personality is impossible. Errors of learning that lead to errors of logic all have their basis in chemistry. And consider the vicious cycle if the converse is true, which it is, of course.

Adrenergia

Adrenaline. In humans, as in all mammals, when the creature is threatened in some way, or there becomes a need for the body to perform at maximum physical capacity, a big surge of adrenaline accomplishes this. You've probably heard of how it prepares the body instinctively for "fight" or "flight", by driving the physiologic mechanisms required for survival in the jungle, from whence we came.

When danger is sensed in the brain by psychological or social threat, this same adrenergic overdrive kicks on. The effects of adrenaline now become physiologic symptoms, and often a person will seek medical attention for what they fear is a serious medical problem on their hands. And since adrenaline has an effect of some kind on nearly all tissues and organs of the body, there is such a great diversity in the symptoms it causes that it has been referred to as "the great imitator" of a multitude of pathologies. It poses a great challenge to primary care physicians, knowing people won't be particularly happy if disease ends up being present, and the doctor told them it was "all in their head". It doesn't help that symptoms caused by adrenaline are the most common thing we see in the ER.

But if the condition in question is felt to be caused by such adrenaline- surging, symptoms should be many and diffuse, reflecting the "whole body" effects of this protective hormone. A discussion regarding these effects on other organ systems may lead to a diagnosis of "nervousness". I like the term "adrenergia".

To wit:

**Respiration -- Adrenaline overdrives the breathing system to prepare for large amounts of gas exchange, providing oxygen to muscles for "the great escape", and to ventilate off the carbon dioxide they generate. With the rib muscles at high tone, the chest feels tight. Adrenaline tells the brain it wants you to breathe deep, and people report they "can't get a full breath". The sharp, painful stab of breathing is the result of trying to stretch these tight muscles. With this hurt usually comes some amount of rib cartilage soreness along the front and lower portions of the chest, the condition of costochondritis (costo: rib; chondro: cartilage).

Without a lot of carbon dioxide being generated, since there is no bear chasing you, and you're not running away, CO_2 levels in the bloodstream fall, resulting in numbness and tingling of the hands and feet, and the face, and the whole body can go numb. Brief, non-dangerous losses of consciousness can also occur, and a person might report they've "been passing out all day long".

**Circulation -- Adrenaline stimulates the heart to beat harder and faster, and can make it beat irregularly. Though not dangerous, such cardiac features add to the fear of the situation, already being triggered in the brain by the chemical itself. Inhibition of blood flow to the skin by adrenaline will result in a certain pale appearance, and the hands and feet may feel cold.

**Digestion -- Adrenaline is adverse to the activities of the gut, in fact embodying the opposite of that vegetative part of us. The pharynx (swallowing mechanism) is stimulated to constrict, giving the well-known symptom of "choking", like when the game is on the line. To many non-athletes, the "something in my throat", or the sense that "it's closing off", is this same physiologic effect. The textbook term is "globis hystericus". It's very common.

The stomach won't relax and take in a pizza when under adrenergic clamp-down, and this results in the symptom of "early satiety", the sense of getting full on a few bites. Vomiting and diarrhea are quite common in heavily nervous individuals, demonstrating the rather irritable nature of this digestive tube.

**Muscles -- They're ready to go with adrenaline, for sure, but when the stimulation is too much, it's more a shake and quiver that can result. That nervous shake.

**Urinary Bladder -- Again, it will expand and fill as a vegetative part of us, but when you're a little anxious the bladder won't relax, making frequent urination a very common complaint. Unfortunately, the adrenergic tensing of the sphincters will make public urination an even more difficult adventure for nervous pinklers.

**Brain -- The mind is quick and shallow under the influence of adrenaline. Its memory and learning are inhibited. There is apprehension and fear, often to mortal levels in the extremes of phobia and panic. Its reflexive behavior may be regretted later. Chronic wakefulness will result in fatigue and irritability, pain in the neck and back, and headaches.

Anxiety

It should come as no surprise that the protective chemistries of the body should "go off" so easily, since it must always be at the ready should a threatening situation arise. Now, with society posing the danger instead of the traditional foes to the human (wild animals, disease, the elements), we can say that it's normal to "be anxious" about certain things in our modern social realm. The symptoms of the consequent adrenergia we learn to recognize, and most of us feel better to deal with things, and get that "something off of your chest".

But for some, "stress" is an overwhelming force, and such people may suffer an incapacitation of some sort (mute-ness and pseudo-vegetation, work or school inhibition), and you could argue they've had a "nervous breakdown". They're carrying a big adrenaline burden, and maybe aren't coping real well. They come in or are brought to the medical system, to seek relief and counsel.

There are people who either don't or can't ventilate their pent-up emotional energies in a healthy and effective manner, and others who are in social situations that are truly dangerous and frightening, worse than most people ever experience. Or both. For them, the chronic drag and harassment of adrenergia results.

If there is no particular stressor, yet the distressing manifestations of adrenaline physiology persist for, say, months, then the condition of "generalized" or "free floating" anxiety is said to exist. Theoretically, a subconscious thought or emotion must be lurking back in the memory somewhere that, an attempt by the mind to sequester it results in a kicking on of defense physiology. Though it would help the brain to expel this thought, such an act would be too embarrassing or shameful, and it is held in. Adrenergia results, and the nervousness becomes chronic.

In psychological and psychiatric practice, a comfortable couch is provided, and the sufferer encouraged to share these thoughts. Hurtful childhood memories such as physical or sexual abuse, exposure to parental "bad behavior" (anger, ambivalence, favoritism), and other perceived character flaws can be discussed and released through the use of such intervention by sensitive, caring professionals.

Still, such therapy may take months and years to benefit some of the more badly hurt patients, and indeed many never recover fully. Furthermore, when stressful situations arise in subsequent periods of life, well-entrenched defense mechanisms can be so easy to fall back on, and fall back over.

Panic Disorder

Panic disorder is characterized by recurrent 15- or 20-minute episodes of sudden and complete adrenalization, coming on, apparently, completely "out-of-the-blue". They can occur from a sound sleep. They also seem to arrive in clusters, playing out over a few weeks where they become more frequent, only to fade off and recur perhaps years later. Like a wave, of behavior. An episode, of episodes. Typically they clearly and classically display the effects of defense physiology.

A Case History: The patient, arriving by ambulance, is a 34 year-old white male who has fallen ill in the workplace. He is shaking, pale, and speaking in halting phrases, and says they called the ambulance because he "couldn't breathe". He says his chest is tight, and when he takes a breath he gets sharp stabs of pain along his sternum. He's wondering if he's having a heart attack.

He's placed in a room where he's hooked up to monitors and oxygen, an IV is started, and a series of heart and lung tests are ordered. But he's feeling a lot better now, is breathing good, and while his chest still feels a little tight, it doesn't hurt as much to breathe as it did before. More history is gathered from him. He says, "Doc, I thought I was going to die." He was sitting at his desk job, on a normal day, and in a split second, all of a sudden these symptoms came over him. He also points out that he feels like he's choking. He's had four spells like this now, over the past few weeks, and he went to his family doctor about it and was told he had a throat and lung virus, and was put on antibiotics and something else, but he's still having these attacks. The other night he woke up like that, scared to death, couldn't breathe, real shaky. He got up and walked around for awhile, and after about an hour or so he was able to get back to sleep. And now this.

His tests all come back normal. He says there's really not any heart disease in his family, and he works out all the time and considers himself to be in good physical condition. Plus, he had spells like this a few years ago, and they did a bunch of heart tests on him, including a stress test, and it was all normal then, too. He asks, "What's the matter with me?"

This is the patient with panic disorder. The episodes of adrenaline surge are playing out in random fashion, peaking over this week or two. His symptoms are so dramatic, and his complaints so bothersome, that people like him are typically admitted to the hospital for observation and further cardiac testing. Some people even argue that you shouldn't make the diagnosis of panic disorder in the ER, because it so much mimics angina and heart pain.

But heart pain is dull and generally radiates somewhere like the arms or jawbones. It doesn't come on so suddenly, doesn't cause a choking sensation, usually doesn't cause painful breathing, and while scary, people usually don't make it a point to say they felt like they were "about to die", which the panic patient so often does.

There should be some form of mood condition in the family of the patient with panic. A bipolar parent, or another of the sped-ups, like alcoholics, gamblers, druggers, or criminals, almost always there will be such evidence of the highly revved mood at the heart of runs of panic.

Question? Why panic? Why is such a free-flowing domination of the brain from such a low center allowed? Again, mood does that. Its neighbor down there in the brainstem supplies the adrenaline, essentially paralleling the mood level.

But still, in the ultimate high mood state, mania, there doesn't seem to be much panic, while in the panic patient, typically a somewhat "hyper" individual, there doesn't seem to be much in the way of delusion. So adrenaline, panic, hyperactivity syndromes, and even mania, are all low-brain phenomena, and are often seen together. (See: The discussion on the tree fractal, in Section One of this guide, called "Simply, Chaos".)

Phobia

Recall the word "neurosis", a condition where a particular behavioral problem has resulted from mental injury, which occurred during the nurturing process. A classic is the one they call phobia.

Phobic neurosis is marked by intense fear and avoidance of certain places and situations that is far beyond the realistic dangers posed by them. When such a circumstance does arise, sudden and complete adrenalization occurs, to the level of panic. "I've got to get out of here!" The attempt to escape will be dramatic and typically embarrassing.

In nearly all cases it was some event that occurred in childhood that so totally frightened a child that a permanent memory was posted in the danger center, and the person cannot alone seem to overcome the phobic stimulus. Did a nurturer somehow drop the ball here? Surely.

It is felt that this disorder can be treated effectively by some form of assisted "systematic desensitization" if the patient is willing, and finds such therapy desirable over simple avoidance. Some, such as social phobia (fear of embarrassment in front of a crowd), are more stubborn. The famous and common "agoraphobia", or fear of incapacitating panic, is a disorder of the patient with severe panic disorder, and is not so much a fear of public places (agoros is Greek for "market") as it is fear of the out-of-control nature of panic itself, and hence it is not a phobia at all.

Minor Depression

Sometimes, you know, life just ain't worth it. The job sucks. The marriage is boring, the kids are driving you nuts. You're getting older. All of a sudden, you just don't seem to be getting much pleasure out of anything. You're getting kinda fat and outta shape. You can't sleep no damn good, you're tired all the time. You had some tests done through work, and found out you're at risk for all sorts of things.

Then you picked up the paper. Now you're mad at all the idiots out there, and you kicked the dog when you got home. You're working late, and all the while you're wondering if maybe you missed a turn somewhere. Perhaps some dream ain't coming through.

You haven't cried have you? Suicidal thoughts? In fact, some people swing to depths of frustration that are simply beyond what can be considered the normal amount of pessimism and gloom. We make a change, or cope with things somehow. But for others, a severe sink into a very real behavioral condition: the depressive episode.

A Case History: The patient is a 36 year-old white female who has come to her family doctor because she feels like she is sick all the time. She's taken several rounds of antibiotics over the past three months for upper respiratory symptoms, but she "can't seem to shake it". She had a lot of trouble getting her breath, and though they sent her to a lung specialist, he couldn't find anything wrong with her, and sent her back to her family doctor.

She reports that she's missed a lot of work recently, and is having headaches "almost every day". She complains that marriage isn't so great, and that her husband is cold and distant. Lately she finds herself crying all the time. On several occasions she has felt so awful that she couldn't get out of bed the whole day.

Minor depression is very often marked by complaint and illness, along with the expression of dark emotions like pain, suffering, hopelessness, and frequently, suicidal ideation. Such episodes are triggered by stressful experiences, and usually are seen in people with marginal personality styles, and questionable coping skills. The depths of this struggle will persist for months.

Treatment with combinations of modern anti-depressants is usually of some value, and the risk of suicide during the early stages of improvement presents another management wrinkle. A family history of such a behavioral phenomenon would be typical.

Reactive Conditions

A common form of minor psychiatry, seen so often by the generalist in doctor's offices and ER's masquerading as complaint, is the variety of behaviors that are brief but real maladaptations to stressors that people encounter, or are befallen. They are minor episodes in that there is no frank delusion or hallucination, like seen in major depression or mania, and because the prognosis is generally favorable. But they are "major" "scenes" for the patient and their family, as measurable social impairment results.

When significant and major events occur in certain people's lives, such as loss (interpersonal, economic), increased responsibility (at work, or at home), illness or misbehavior in family members, shortcomings in terms of coping styles may show as a minor episode of misperformance by the mind manifest as a "mood disturbance." This run of difficulty has been referred to as "demoralization", an aberrant behavior in a person who has generally fared well, but has taken a hit that he or she simply cannot handle, and reacts in a bizarre fashion.

Such "adjustment disorder" is common in children and adolescents, but can happen at any age. It will usually occur within a few months following the stressor, or trigger. Often a depressive episode like the one illustrated above will occur, the resulting inhibition marked by lots of days missed at school or in the workplace. There is seeking of medical attention, and passiveness.

Sometimes the opposite occurs. Agitation, irritability, and a load of adrenergia may predominate, the condition of anxiety. There may be an inappropriate mixing of emotions, such as laughing at misfortune, or perhaps a conduct disorder routine. Or even frank ambivalence. Whatever the case, social impairment will evidence that an episode of behavior is underway.

We use the term "self esteem" to speak of that part of our personality that values dignity, decency, and borders regarding the extremes of behavior. Perhaps we were firmly taught to follow certain rules, avoiding the self-defeat of maladaptive schemes. Some of our population will simply "shut down" their social instincts in times of assault or accident, adopting behavior that everyone agrees is abnormal.

Case History: The patient is a 35 year-old white male with the chief complaint of headaches. He rarely has gotten headaches until a few weeks ago, and he describes the classic form of "tension", or muscular headaches. He agrees that he has been restless for the past month, that he has "no appetite", complains of fatigue and irritability, and thinks his memory is failing him. He says he's getting worse, his mind is racing, and last night he didn't sleep at all. He missed work at his factory job and they're concerned about him. Extensive testing for headache causes and other health conditions done by his family doctor have been normal, but he and his supportive family are sure something's wrong with him. He denies drug or alcohol use.

He grew up fairly normal, and after high school he went to work. He has been reliable and decent. But things at work have been "stressful", and his dad, a smoker, was just diagnosed with lung cancer a few months ago. Other than that, no major events in his life are identified.

Very often, these are the types of minor elevations of the mood, the anxious and irritable ones, that we see accounting for a multitude of symptoms attributable to a sustained rise in defense hormonal physiology. The brain is sensing danger, and is wearing the body out with its adrenaline. The wakefulness of this alerting chemistry is affecting sleep depth and quality, resulting in fatigue, muscular headaches, and the poor performance of a poorly rested mind. The gut and appetite inhibition is resulting in poor caloric intake, exacerbating these symptoms. It has been triggered by stressors real and/or imagined, and he has no insight, and is skeptical of your evaluation and conclusion. All he and his wife know is he feels terrible.

This bout with "adrenergia" will go on for several weeks, maybe some months, and he will return to his normal, relatively uncomplicated behavioral baseline.

In some prone people, especially those with more dysfunctional upbringings, such minor mood elevations will be punctuated by a series of panic episodes, usually over a few weeks in the middle of the experience.

Case History: The patient is a 28-year-old female who says she is "depressed". She's been feeling this way for the past few months, and recognizes that she doesn't seem to be getting any better with the Zoloft her doctor has prescribed for her. She is not suicidal and has no psychotic behavior (hearing voices, etc.). She has been to the medical system several times, despite her impression that no one seems to be able to pull her out of this.

It seemed to start when she got a promotion at work a few months ago, where she is a critical, hard-working employee (she manages a fast-food joint). She has two young kids, and her husband is a loser and left a few years ago. Her father was an alcoholic and she barely knew him. Her mom was loving and decent. She denies drug or alcohol use.

The problem today is that she doesn't feel like she can go to work. When she has gone in over the past few weeks she gets real nervous and has been found in her office crying, and thinks she's been having panic episodes.

Underlying all of this struggle for her is a sense that there is too much burden for her to bear. When more responsibility was asked of her in the workplace, she soon became symptomatic with depression symptoms, sick role playing, and work inhibition. The combination of this, lack of companion support, financial woes, and the responsibilities of her needy young kids is too much. She just…. can't do it.

It is assumed that people with such esteem problems suffered nurturing failure, and that probably I.Q. (logic, memory integration) is playing a major role. Her mate selection, predictably, was poor. While "the going" has gotten tough (well, tougher than usual), rather than adapt to the new atmosphere and environment, a sense of doubt and danger became pervasive, and for her these physical symptoms and sick-role playing embody the escape. The enablement by good-doers is obvious.

Never forget that the lower rungs of society are there, generally, because of a relatively much higher amount of minor and major mental problems in them, their siblings and parents, and their acquaintances as well. It is the reality, indeed the essence, of life in the lower "class".

Thymic Disorders

There are some people who display chronic mood levels that are either above or below what might be considered "normal", and we say they have "thymic" disorders. Presumably the word "thymic" comes from a belief that it is somehow the gland in the chest called the thymus that is responsible for mood levels, but apparently this is not the case. Mood is most definitely a nervous system thing, and the thymus is an immune system gland. Interestingly though, mood state, mental wellness, and immune performance are tied together, we know that.

Recall that "mood" is the relative energy, and level of thought, generated in the nervous system. When the mood is high, a person feels good physically, feels good about his or her performances and achievements, and is optimistic. When low, all of the opposites apply. Ailment, complaint, pessimism. Mood, then, is a sustained pattern of emotion.

Momentary reactions of everyday inter-actions, on the other hand, is what is called "affect". Some people may come across to you as flat and cold in their emotional level, while others wear you out with their odd, bizarre, or silly and flippant affect. When you talk to someone, you're immediately noticing how the discourse is going, and sense, hopefully, affect.

While mood and affect are inextricably tied, they are entirely different and separate. For the most part, mood can be considered more the physical being, while affect is a mental and social thing. There are conditions that seem to be mood-only phenomena, and there are affective syndromes that are quite distinct from mood disorders.

In thymia, the mood is "set" at one rather far end of normal, and while it can be expected to fluctuate some like all things, it does not show episodic qualities. Since there is no hardcore delusion or hallucination, these people aren't what you could call "crazy".

If your mood is somewhat continuously turned up, and you seem to always have "big plans" and exhibit limitless energy, sleep only a handful of hours a night, while at times showing impatience and irritability, then we say you have a "hyperthymic" mood. As a steady-state performance, hyperthymics rarely or briefly swing downward. Substance utilization and abuse is common in these heavily curious and fascinated individuals.

Amazing or annoying, in fact a number of famous humans throughout history have shown this temperament. Chronic, and just short of mania, it is a condition sometimes called "hypo-mania". In the absence of nurturing problems it is best considered a gift.

Then there is the cruel opposite, the chronic poor mood state that we call "dys-thymia". People with this temperament are typically pessimistic, introverted, and tend to voice their complaints. They are preoccupied with failure and shortcomings, in themselves and in others. As the "downers" of the world, they will complain a lot and try to bring you down there with them. Their frequent sachets with the negative vibe casts the "Oh, no, not again" shadow over their everyday life. Usually, the next shoe does drop, as they suffer life, rather than live it. When their low, self-defeating mood level dips a little, suicidal risk will become significant.

Treatment of such "characterological" depression with "mood elevators" (antidepressants) is often helpful, but the low of the low ebb will at times necessitate more aggressive intervention. The condition is life-long.

If you cycle from high as a hyperthymic to low as the dysthymics, and do it on a few-weeks-to a-month rhythm, then they call you "cyclo-thymic", and you can probably recall an occasional-such mercurial person from your journey. As with all thymic phenomena, events matter, and will serve to trigger swings.

Major Mood Disturbances

Elevation of the mood (mania), or depression, to the point of delusion and or hallucination, for weeks or months, is manifest in the behavioral phenomenon we call a "major mood disturbance." Generally, if you display a manic episode, we figure a major depression will also be suffered at some point, and hence the term "bi-polar affective disorder". Depression alone is called Unipolar Major Depression. With these conditions, there is complete remission between episodes.

In mania there is a parallel rise in the mood along with rises in vital sense and self-attitude, as illustrated below:

Typical manic case history: The patient is a male (the incidence is about equal), early twenties, and his buddies dragged him in to "get help for his nerves". They hand me some papers he wrote that have all sorts of things written in different directions, with arrows here, and underlining there, and circles and stars. Things in the margins.

He says there's "nothing the matter with him", and can't understand why there's something wrong with calling people all hours of the night. Though he hasn't slept in days he says he "feels great". He's got a fiancé, is thinking about moving to Florida, and just quit his job. He's spent all his money, and run up his credit cards, but is certain his plans for success will be realized, and that he will be able to pay everything back. His friends say he's "getting worse", and hasn't seemed exactly right for months.

Manic episodes are among the most dramatic of all behavioral phenomena. The talking is essentially nonstop, as a steady stream of mood exuberance is spilled forth. The travel from topic to topic is referred to as the "flight of ideas", and the relative flippancy of these topics strung together as "loose association". In extreme cases delusions of paranoia and persecution, and this high mood state, can present significant danger to society, and police and their healthcare cohorts.

Case history: Unipolar Major Depression The patient is a female, and she's fifty. She's been normal her whole life, but something is clearly wrong. She's been going to the medical system a lot lately, and she's frustrated that they can't find anything wrong with her. She's tried medicines they've given her, and thinks they are making her worse. She finds herself crying at times, and for the past several weeks has lost interest in just about everything, has lost her appetite, and seems to be sleeping poorly. She's refusing to bathe now.

Today, she was brought in because she called her husband to tell him there are large bugs crawling all over the walls of her kitchen, but when she called her neighbor in a panic, no one else could see them.

She has "major" depression. Typically there is vegetation, withdrawal from social interaction, delusions of paranoia and persecution, and visual and auditory hallucination.

Fascinated, I tell her she needs to be in the hospital to "get better", and I reassure her family. Next she looks up at me and says, "But what about the bugs?"

In mania and major depression, the behavior as perceived by others is decidedly abnormal, and their inappropriate social actions make them qualify as major affective syndromes. There is a parallel rise and fall in their mood levels as well, making the manic feel too good to sleep, and the depressed counterpart feel too bad to come out of the house.

It is worth mentioning that a lot of people with dramatic personalities offer wrongful insight with the old "I'm bipolar". Typical of their exploitative nature, such personality disorder people are attempting to both draw on the general likeability of most mild manics (they are much less likely to exploit others), while exonerating their self-mindedness as a "chemical imbalance".

Such false and phony insight underscores the potential depths of dependent behavior, and dramatic personality.

Somatoform Disorders

So far, we've talked about conditions that are more or less common and regular. But there are those humans out there, who so mangle and botch nurture that their offspring pull positively outrageous behavior that captivates gatherings, cities, and at times, whole countries. Minds, so broken, delve into mild and modest delusion as a steady-state MO when interpreting events and actions of others, and live a tormented, weird-out of an existence as a result. At times, big shows.

A good example is what we call somatoform behavior. There are a handful of severe, chronic, and relapsing behavioral conditions where the patient experiences ailment and no medical abnormality can be identified. They are in the "form" of a bodily disease (somato: body), but in fact the problem is not a pathology, but a major mental ailment. It should not surprise anyone that they turn out to be an expression of a severe emotional conflict.

These are very significant and difficult psychiatric phenomena, and the sense of suffering on the part of the patient is large. Moreover, they are among the most stubborn and refractory problems to manage and treat.

The three most notable somatoform behaviors are somatization disorder, conversion hysteria, and hypochondriasis, with the recently added "body dysmorphic disorder" rounding out the group.

Common to all of them are extreme preoccupation with symptoms, medical-attention seeking, and the inability to be reassured that "nothing's wrong". When the suggestion is made that the symptoms could be behavioral, the patient can be expected to react with disbelief, anger, rage, and resentment, and will then move on to a new doctor who might take their complex charade seriously. And while they generally feel dissatisfied with the medical system, they're accessing it on regular occasion. Hence, it is an illness behavior.

Recall the "vital sense" as a basic part of functioning. Most people feel like they feel OK, while some, like the manic patient, feel "too good", and can do all these things. The depressed mood patient feels ill, weak, and hurts. But in the somatoform disorder patient it is an intense fear of, and preoccupation with, illness, and an inability to be reasoned with.

These conditions, with their roots in nurturing injury, appear to be the use of illness to mete out a rage and anger created by the mind's mistreatment. They don't like their body, and don't trust it. They see it as diseased, and can't be convinced otherwise. All of this may indicate that emotional and mental hurt somehow interferes with a stage of development as simple as bonding of the mind and body, opening the door to the social realm where you aren't so focused on yourself, indeed disease of the self, the essence of somatoformia.

*Somatization Disorder: In the patient with somatization disorder there is an incredibly vast array of symptoms that seem to include all the major organ systems of the body, and they are all occurring at the same time. From headaches and fatigue, to menstrual problems and digestive complaints, to breathing difficulties, to concurrent "pseudo-neurological" symptoms (dizziness, weakness, numbness), these patients, usually women, appear to be saying that their brains are interpreting nervous system messages from these organ systems in some form of dysphoric way. While most people's bowel activity, for example, goes on below the conscious level, in somatization disorder the sensations are not only making it to there, but they are perceived by the mind as misery and a sense of suffering. "It's killing me, doctor. Can I have....something for it? Please, help me."

The book says that when there are four or more organ systems on the complaint list, we are likely to conclude that "somatization" is occurring.

And, interestingly, these patients have a very difficult time explaining their symptoms, and the trained clinician soon recognizes, through expressionism and body english, that the complaint has its basis in behavioral pathology.

There is considerable morbidity in this life-long somatoform condition, and doctor-shopping, frequent accessing of the medical system, multiple extensive workups, and unnecessary surgical procedures are the norm.

It turns out that first degree male relatives of females with somatization disorder show an increased incidence of alcoholism and antisocial personality. The old nature vs. nurture argument is begged here: are the men anti-social because of a mother who cannot nurture due to her perceived physical ailments, or is there a connection between somatization, a sense of misery and chronic anger that results, depression, and anti-social coping mechanisms?

*Conversion Hysteria: Formerly known as "Briquet's Syndrome", this disorder is marked by a patient's perceived sudden loss of function of a part of the body. Typical conversion experiences include loss of sight, inability to walk or move an extremity, to swallow, or to urinate. A feigned seizure is another conversion phenomenon. Again, the trained clinician soon recognizes through physical diagnosis that a somatoform condition is at hand.

Generally these sudden-onset behavioral pathologies can be linked to stressful events in the patient's lives. One or more very concerned (doting) family members will be playing their role as legitimizers to what is essentially a performance utilized to cope with a painful experience, or as a way to demand interaction from a loved one whose attention they do not or cannot get by a more reasonable and healthy means. Without the accomplice, conversion hysteria would not be something this brain would try to pull off. In fact, it is these relatives who present the biggest challenge to the clinician who is suggesting that a somatoform explanation applies, while the patient can be seen, out of the corner of the eye, to exhibit an inattention to the problem famously referred to as "la belle indifference".

When the patient is challenged directly, the ego shattering that results is manifested, as usual, by anger and resentment, and disbelief on the part of the sufferer, further solidifying the diagnosis. Such distraction generally "cures" the "ailment".

These brief bouts with somatoform behavior are somewhat common in children of adolescent age, but certainly occur in adults as well, most of whom have previously been diagnosed with another form of behavioral disorder.

*Hypochondriasis: Occurring with equal frequencies in both sexes, the mindset of hypochondriasis is marked by exaggeration of a physical finding, morbid fascination that disease is present, and refractory responses to even spirited attempts by the clinician to convince the patient that all is well. This famous and infamous somatoform condition causes significant chronic suffering in its victims, who will generally move from one area of the body to another over the years. While the patient can and will provide signs and symptoms in great detail, they are unrecognizable to the trained clinician.

As a condition that crosses all socio-economic classes hypochondriasis is seen frequently in medical practice. Presenting in middle age, it shows a chronic fluctuating course, with perhaps only 5% of sufferers recovering fully. If it is part of a depressive illness it can be particularly difficult to manage. More often, these are patients with compulsive personality characteristics, who respond poorly to criticism, and struggle to express emotion and feeling.

Aldrich (C.K.) delineated the following understanding of hypochondriasis and other somatoform disorders:

1) The symptoms allow a "break" from the normal functioning of life, with the sick-role-playing serving as a diversion and coping mechanism.
2) There was nurturing failure in how childhood illness was handled by the parents, such as excessive emphasis on minor ailments and insistence on sick role playing.
3) The ability of the medical system to provide "attention" (tests, diagnoses, medications) to these adults who have no other way of finding such missing nurture,
4) The continued nature of complaint and concern serve to keep the gig going.
5) These patient's are unaware of why they are doing what they are doing.

Consequently, the following can be expected: Reassurance is seen as rejection. Extensive testing, etc., jibes with the patient's concern for illness, thereby reinforcing the behavioral pathology. New symptoms can be expected to emerge continually.

Obviously, managing such patients requires a clinician who is caring, limits and resists the patient's desires for extensive evaluation and referral to specialists, and listens to the newest round of symptoms with what appears to be genuine concern. And he'd better, because even hypochondriacs die of something.

*Body Dysmorphic Disorder: New to the list of somatoform mental illness is this one, characterized by a morbid fascination with some body part that is considered grotesque and displeasing, while others do not see it that way. A typical example is the patient who feels their nose is a revolting structure, and undergoes multiple nose jobs and can never be satisfied. Or they feel they are obese when they aren't.

Normally, this type of behavior is generated by nurturing failure. Somewhere, somehow, the developing mind was forced to concentrate too much on body image or concepts of beauty. As a result, they cannot see their body in what might be considered a "correct" interpretation of the reflection in the mirror.

Overly controlling, adoring parents who treat their child as this self-handed "prize" of reproduction, have constructed what has been called the "golden cage" around a child who will later develop anorexia nervosa, the most well-known condition of this wrong-behavior category we call "Body Dysmorphic Disorder".

Dissociation

Among the neuroses is the phenomenon of dissociation. Simply stated, the failure to properly integrate occurrences and sequences of events into the memory is called dissociative disorder. Suddenly, in dissociation, it's as if all events occurring in the present consciousness are being sent to some new "memory bank", in what is essentially a new identity.

There is a separate section of this guide that explores the concept of identity, and how memory storage and recollection affects behavior in general. For now, a brief run-down of these rare and astounding behaviors:

In "dissociative amnesia", a person has "forgotten" important personal information, such as who they are, where they were, and what they did, for what is typically a few hours to a few days or weeks. Sometimes only parts of these time frames are lost, and the patient will recognize that something is wrong. Others will only remember when they are clearly shown something they did that they cannot recall.

Dissociative amnesia is most commonly seen in young adults whose minds utilize this "trick" in an effort to deal with an unpleasant memory, such as sexual abuse or horrendous tragedy. Often the memory will return in a patchwork fashion many years after it has occurred. "A blackout". Such awakenings are usually associated with embarrassment and anxiety, and ego defense such as mutilation and suicidal gesturing (rejection of the identity).

Some will never recall the events of their traumatic past. Others will, for some reason, have better overall psychological adjustment abilities, and better periods in their lives will allow them to reconstruct their past and move forward. Continually sequestered memories, however, are likely to result in bouts with anxiety and depression.

The "dissociative fugue" state is a disorder marked by complete loss of memory of one's entire past, with the creation of a new identity and a sudden, unexpected change in address, usually far from home. Though unexpected or "unintentional", the performance is purposeful.

Fugue states differ from simple amnestic experiences in that the person is utilizing (subconsciously, of course) this identity-morphing as a form of malingering, or as a way of escaping certain responsibilities, or as an attempt to absolve the self from things like criminal behavior. And this is probably aided by change of cities and all the reminders of that old self. There is a term "geographic instability" that applies here.

Generally, fugues are brief and self-limited. Eventually the person becomes confused over his or her identity, and suddenly becomes aware of who they are and have been. With both fugue and dissociative amnesia, hypnosis and drug-assisted interviews may be helpful in recovering the real pasts for these persons.

In fact, it appears that people who suffer dissociative psychopathologies are also easily hypnotized, in general, and may have adverse reactions to that rather present and available hypnotism drug, alcohol.

"Dissociative Identity Disorder" is the form of dissociation that was previously referred to as the somewhat well-known condition of multiple personality. In this fascinating behavioral disorder there are at least two concurrent personalities coexisting in one individual. One identity may know some, but not all, things that the other knows, and vice- versa. Often they are oblivious to one another.

As opposed to the more benign fugues and amnestic states, dissociative identity disorder is chronic and severe, and is associated with the highest incidence of suicide of all behavioral disorders. It is more common than previously thought, accounting for 3 to 4 % of acute psychiatric emergencies, many triggered by substance abuse.

Finally, there is the phenomenon known as "depersonalization disorder". In this form of dissociation there is a recurrent sense of being "outside the body" observing the self, which may appear as distorted and unreal. They may suffer a distressing sense that they were paralyzed or were being inhibited in some way. Memory is fragmented, and when they "snap out of it" some hours later, they will note that they "lost some time", elapsed under their nose while they were trapped in this lost journey. Some suffer the condition in a mild form, while others will ruminate about these experiences, fear they are "going crazy", and suffer significant impairment and disability.

The phenomenon is probably much more common than was previously thought. Some say as much as 1% of the population are capable of such brief weird-outs. Usually the trigger is a heavy duty stressor of some kind, like getting on an airplane, or social duress such as loss and bereavement. Perhaps some minds attempt to protect themselves in this manner, to the chagrin of the conscious identity.

For the most part, doctors of primary care are unlikely to understand these symptoms that a patient cannot hope to describe. "My arms were so heavy, I couldn't move." "I think I passed out or something". Perplexed so frequently by unusual descriptions of malady, the clinician soon wonders if a behavior disturbance is at the root if it.

Identity

Identity is developed in childhood, especially during the years from about 6 until puberty. In people who are failed by their nurturers, and experience severe neglect and profound physical or sexual abuse, or are chronically ill during this crucial period of development, they may grow up to exhibit this dissociative phenomenon. It is possible that their identity does not develop correctly or fully, and perhaps that is why they can simply abandon it and its memory storage patterns, allowing them to uncouple the integration of life's events as they are unfolding. The formation of an entirely different identity is used by the brain as a way of dealing with stressful new events.

Hastily manufactured identities will strike the observer as bizarre, since typically they are fragmented and underdeveloped. They may resemble anxiety or depressive states, or may come to medical attention due to autonomic hyperactivity and the associated symptomatology. A diagnosis of personality disorder, psychosis, or psychoactive drug effect may seem appropriate.

We will suspect the use of multiple identity when fluctuations in level of personality performance is noted. From high functioning to disability; dramatic swings in mood; time losses and amnesia; depersonalization (feeling detached or removed from the self) and derealization (forgetting familiar things, people, and places). And, of course, the essence of dissociation, failure of memory.

Personality

Now, and finally, personality. We've talked about moods and episodes, and anxiety and depression. And goon-outs and ego defense, and poor self esteem, and the wacky extremes of somatoformia and dissociation. But the day-in and day-out you, if it's off the mark of expected social, logical, and considerate interaction, then we say you have a "personality disorder".

The life-long attitudes, beliefs, coping mechanisms and relative insight constitute the social capabilities of a human, and we call this the personality. With a number of components, or dimensions, that interplay in the resultant behavior, a balance of these "traits" is needed to allow for some kind of "normal" performance, by which we can compare those members of society who draw our attention for one reason or another. With the level of mood taken into consideration, along with IQ and logic, an experience called "growing up" will shape these factors into the end-resulting unique creature we all become.

For the most part, when we're full grown at around 20, these personality traits will tend to "identify" the behavior of a given individual, and can be expected to change little over the life-span. And, if one of these traits is significantly out of balance, then it can be expected to be that way forever. In personality disorder, one of these elements of self is wagging and dominating the behavior.

It's easy to say that personality disorder is some kind of euphemism for the losers, weirdos, and asses of the world, but what is really important is to recognize that a person has such a condition, and knowing that they will never change and cannot change. Thus, there are three simple rules in dealing with people so-afflicted: avoidance, avoidance, avoidance.

The DSM IV classifies the 12 or so maladaptive personalities into "clusters", and there are three. There is Cluster A, the "odd/eccentric" types, Cluster B, the "dramatic" individuals, and Cluster C, the "anxious/ inhibited" bunch. Such grouping implies that similarities exist, and these are evident in how the very basic levels of understanding oneself are essential to the achievement of decent social capacities.

In "personality disorder", it is helpful to identify the behavioral pattern, and through further information of the person's background, to understand what may've happened to cause one of the dimensional personality "traits" to be so out of balance. Generally, the cause is easy to pin down.

The "Odd/Eccentric" Cluster

Individuals we consider "oddballs" suffer from a mild disorder of logic. Their interpretations of events around them, and of the aspects of social interaction, are inaccurate and lead to actions and reactions that others cannot understand, since they don't think that way. Consequently, people avoid them from pre-school age onward, giving them the feeling that the world is a cold and cruel place. The effect of such long-term isolation worsens the situation.

In "paranoid personality" the person behaves with a general suspicion as to the the motives and intents of the actions of others. An old term "querulous" refers to the tendency of some to basically question everything that goes on, from fixed sporting event outcomes to the activities of the people next door. When they do succeed in a relationship they will generally become jealous and controlling, and may react with rage if their suspicions are confirmed. Their cold nature will be expected to drive off would-be suitors in general, perhaps bringing only the more dysfunctional members of the opposite sex their way.

In cases where the logic process is more impaired, "schizoid" and "schizotypal" personality become the diagnosis. Rather than paranoid, these persons are emotionally detached and withdrawn into their own world of fantasy, daydreaming, and theorizing. "Schizoid" persons fear intimacy and keep to themselves, while the "schizotypal" personality is willing to share his odd and magical thinking with the world. Manifesto people like the Unabomber and the Oklahoma City bomber are classic schizotypals. They have "magic" in their minds, thinking certain actions will have illogical outcomes.

The thought and logic disorder in these people is felt to be a muted expression of the genes that cause schizophrenia, and hence the term. Presumably the parents themselves must have a collection of genes between them that interfere with the basic logic and interpretive abilities of the mind. Between all involved, nurture was unsuccessful in getting the child to a proper and logical evaluation of social unfoldings.

The "Dramatic" Cluster

Famously, there are members of society whose behavior is marked mainly by a dramatic performance. All day, every day, drama. A high-mood set of disorders, humans in the so-called "Dramatic Cluster" will certainly get your attention.

Along with this drama, these personality disorders have in common an extreme sense of self, or an extreme focus on the self.

The behavior of others, and of society in general, is interpreted and manipulated by these individuals in a life-long effort to serve the self. An "id" on the loose, if you will. So, in fact, they are delusional phenomena.

Remember, avoidance, avoidance, avoidance, especially with these folks. There are four: The anti-social personality, the borderline personality, the narcissists, and the histrionic (hysterical) personality. Anti-social personality is likely the male mirror of the borderliners, who are traditionally women. And narcissism, like Narcissis, is a guy thing, where hysteria, of course, is girl behavior. So there is, perhaps, this subdivision of the dramatic cluster, indicating that in these two pairs of personality disorders there is a common trait we can identify as problematic, but please understand these are fairly discrete personalities and they are all present in both sexes.

Diagnosis: Sociopath. Seen this guy? Looking at you out of the top of his eyes, and talking to you out of the bottom of his mouth? RUN! He's the most dangerous character in society.

If he catches your eye he might get you, because he can feel your weakness. He knows you're a-scared of him. He's happy to lie, cheat, steal, and be on the move. He's never held a job. He might be in another town tomorrow. Bet your life he smokes.

Perhaps the key feature of this behavior pattern is the keen ability of their reasoning to lay all blame for all things elsewhere. They can use this nifty maneuver to justify almost any exploitative act you can imagine, so they are constantly on the take. In this way they are dependent on others, as a supply for their needs. And as if a created behavioral phenomenon, there is drama, of course. Throw in a little alcohol and Look Out!

Signs of nurturing failure are evident early. From delays in reaching developmental milestones, to fighting with peers in grammar school, to thieving and lying in adolescence. Running away, hanging with people who "flaunt" their antisocial beliefs, substance abuse, and brushes with the law are characteristic behaviors. Alias is not uncommon.

Homemade tattoos, ubiquitous in the sociopath, could be simple dysphoric bloodletting that somehow alleviates a chronic inner pain. There appears to be some form of comfort in these petite and controlled acts of carnage. Depression, felt to play a significant role of the plight of this mental illness, shows you the exploration of negative thought, in fact an obsession with such thought. More reasonable parts of the mind may perhaps reject this identity, and display this "turning against the self" ego defense mechanism. Whatever the case, home-made tattoos are an alerting signal for you to ignore if you dare.

It is felt that the remorseless, amoral, self-minded components of this behavioral disorder are the result a complete failure to establish any amount of nurturing emotional ties. Some combination of bottle feeding, ineffectual touching and stimulation, and neglect in critical periods of perceived needs of infancy and early childhood leads to a child with no concept of social behavior, or what it means to be a member of a society.

While it's easy to blame mom, some kids, maybe, just can't be nurtured. And really, doesn't the whole village have to fail you? Can't somebody put a halt to this big, long "acting out", baby crying syndrome, before it's too late? Possibly, and possibly not. It was discussed earlier how women with the dramatic somatoform disorder "somatization" (chronic sense of multi-system ailment and suffering, without evidence of disease) have a high incidence of antisocial males in their immediate family tree. Moms, with their juxtaposition in the very earliest and most critical stages of nurture, offer society's most important ally in the generation of a healthy populace. So remember this: when dads fail, we get misfits of society; when moms fail, we get monsters of society.

Some sociopaths give up the gig in midlife, and go on to live out a few fairly decent decades. But their flare for the dramatic will never disappear, allowing the keen observer to peer through the rough, ingratiating exterior at the chronically troubled character within. Some have suggested that this apparent change of heart is from not wanting to go "back in the joint."

Regardless, most live shorter lives than their cohorts.

Diagnosis: Borderliner. Borderline personality disorder is characterized by exteme emotional instability, profound dependency, and a morbid fear of abandonment and separation. These people have intense, stormy relationships, filled with dramatics and dazzling manipulative maneuvering which all share the same goal: to foist a demanding dependency on someone. With perceived abandonment they may "turn against the self". Repeated suicidal gesturing, even over trivial matters beginning in adolescence, is typical of the patient with borderline personality disorder.

At times their emotional and mood levels can become so severe that they can break from reality with psychotic behavior (delusion, hallucination). Somatoform disorders are not uncommon in the patient with borderline personality. It is this closeness to psychosis that brings the name "borderline". Remember that. These people are half crazy.

There is a theory that, somewhere very early in the nurturing process there was an extreme sense of loss experienced by the child, or possibly by the mother, perhaps around age two. This intense preoccupation and fear of loss results in a type of social behavior that relies too heavily on others. Such profound dependency, and pure lack of a sense that one should look after oneself, leads to chronically failed social relationships, drawing the irony of the situation full circle: to want so desperately, yet to wind up so desperately alone, the morbidity so feared in the first place.

We notice that to the borderliner everything seems to be black and white, as if either "all good" or "all bad" (the so-called "splitting" phenomenon), and this may make them love you or hate you, and nowhere in-between. You'll feel the same way about them. For most though, the phoneybaloneyhood of the borderliner makes them "the people you love to hate", and you avoid them, avoid them, avoid them.

This difficulty with appreciating the "great gray-zone of reality" may be at the center of this pervasive behavioral disability, like a series of doors never opened. What's locked them? Why can't we unlock them?

Unfortunately, about 1% of the population is felt to have a borderline personality. Few achieve very high, and those who do are a disaster for the workplace. The flare and charm of their dramatic performance may hook the unsuspecting lonely folk, who soon find they're very sorry.

Diagnosis: Narcissistic Personality Disorder. Typically a man, the "narcissist" cannot get enough of himself. He is grandiose and feels superior to others. He has a deep need to be desired, and will tolerate criticism and failure very poorly. He may achieve high or not. He is incapable of feeling remorse of any kind.

And, since he is in this cluster, he is dramatic. Persons with this personality typically have stormy romances, generally centered around some exploitation on the part of the diagnosee. Delusions of grandiosity, exaggeration of personal social standing, and what could be called a form of "charm" (a lie), inflict on their unwitting victims a pattern of self-serving, control-based behavior.

Exaggeration of the self in this manner is an indication of a loss of esteem in childhood, like perhaps when another child is born, and parental attention is diverted. More likely, a parent gave them the sense that "nothing they ever did was good enough". Still, such an impression may be a misinterpretation to some degree by the child, but this impression becomes reality. Was an opportunity missed?

Diagnosis: Histrionic Personality. Persons with this disorder are in constant need of attention, are extremely conscious of looks, and show a dramatic behavior. Essentially every aspect of their behavior is some form of over-reaction. Inappropriate displays of emotion, along with flippant, superficial interpersonal relationships, are characteristic behaviors. Mostly, they are female.

There is a theory that some form of daughter/daddy de-tachment problem is at the root of this chronic behavioral disturbance, as if going from the dramatic, silly-giggly girl of youth, to a mature woman and all the sexual development result in a created behavior, where a shield of drama is covering a confused, indecisive quasi-mature mind. We know how important solid fatherhood is in the development of girls. A question could follow: is the male counterpart of such dramatics from mommy failure? Such oppositry would immediately increase the likelihood of just such a connection.

At any rate, the wear of such sustained performance can be expected to bring swings of depression, mood disturbance, and somatoform experiences.

Generally, some dumb guy will be thinking he's hit it big with this bubbly babe, only to find he is terribly mistaken.

Cluster C : Anxious/Inhibited Personalities

Diagnosis: Dependent Personality. In this peculiar behavioral disorder a person has managed to defer all of life's significant decisions to someone else. Typically, they will be attracted to someone with a conjugal need situation, such as a "control freak". Notice that both require the other to pull off their exploitative approach to survival.

Passive-dependent personality people presumably notice that there are plenty of others who will be a party to your pitifulness. Nuns and social workers will be happy to feel good for having "given" to one so needy. If you so desire, they will literally pick you up, and bathe and feed you.

Again, whatever works. The brain will keep doing it, and why wouldn't it? Passive "style", with all its far-out flavors, is poor behavior, and such underachievement brings to these people a stream of negative energy. It is this lack of insight, and the consequences of such presumptive and allowant behavior that is the essence of it. Because surely, it is a life of suffering.

Sick roles, disability mindset, welfarism, it is these conclusions that propagate the sense of dependence. If this were not accompanied the pain-proneness, depression doldromhood, and failed relationships, perhaps the gig would be more pleasant. The saying that "10% of the people have 90% of the problems" refers in large part to people who are generally passive and "unfortunate". They may even see themselves as warriors in the battle against the big enemies and bad guys and girls of society, as the "long-sufferers", while careful observation reveals who's really to blame.

It is felt that people with passive dependency suffered severe undermining during the nurturing process. This seems obvious, and is generally true. But what is intriguing about passive people is that in many ways they are exerting far more energy doing the nothing that they do, usually more than it would take to just, well, be assertive a little bit. There is something in having others do and decide for you, a closeness perhaps, an intimacy that is sought. A comfort needed. And despite all the hurts and heartaches, they stay at it, even with drama if needed. At some point you'd want to raise an IQ question, but that misses the point. People do what feels right, not necessarily what seems "right".

Chronic Benign Intractible Pain Syndrome

The depths of passive-dependent personality can be explored in the patient with Chronic Benign Intractible Pain Syndrome. In this condition, there is repeated visiting of the medical system with the chief complaint of pain (more than two months), extensive evaluations have failed to reveal any pathology, and the patient is seeking narcotics. It is a problem created by modern society, which engenders it with its parent/child separation pressures, enables it with its extravagant affluence, and promotes it with a liberal mentality, and legal skullduggery.

It is impossible to forget that, in these cases, it is the patient who ends up with all the pain. Accompanying this profoundly negative emotion are the others, including blame, obsession (with pain), disillusionment and depression, anger and resentment, and eventually, complete work (contribution) inhibition. The look in the eye of these social dropouts, playing out their drama, bears a striking resemblance to its real-deal bad cousin, the sociopath.

Back in 1959 G.L. Engel described something he called the "pain-prone" patient. He noted that in many such individuals there was a history of physical or sexual abuse or extreme parental deprivation. This error in development presumably had set up thought patterns that inhibited the brain's ability to grow into more advanced levels of learning and social behavior. Such unresolved conflicts are expressed later in life as a chronic pain syndrome. He found the following features:

a) Unconscious yearning to be cared for and cared about, the essence of passive-dependent personality; this is defended against with the "pseudo-independent facade" (I'm a fighter!" "I've got a high pain threshold".)

b) The utilization of pain in the place of other feelings to communicate emotional needs, such as the desire for love and companionship.

c) Depressive symptoms and low mood state, which the patient feels is caused by the chronic pain, but which probably preceded it.
d) Poor coping skills regarding stressors in family, the workplace, etc.

The fact that narcotics are a part of this puzzle is particularly interesting. Undoubtedly it is the drug's euphoric effect, not its pain reliever effect, that hooks these people, because it isn't so much that they feel less pain when they're taking them, it's that they feel so much better, like they feel "normal". Figure that this is because they are "dysphoric" in a baseline sort of way, and these drugs bring them up to where everyone else is. Once exposed to them, their chance of wanting to make narcosis a part of their life can be immediate.

They'd be pretty good drugs for such behaviorally afflicted individuals if not for the fact that the body will adjust to their effects, what is termed "narcotic tolerance". A single pill's effect might last six hours, so to become addicted to them you either take them four or five times a day, or use long-acting preparations. Within a few weeks, the adjustment by the body takes place. Now, unless more narcotic is continually supplied, withdrawal symptoms will begin, within hours.

Narcotic withdrawal syndrome is not a life-threatening condition. Regional pain and misery, a sense of suffering, and the poor thoughts and emotions of dysphoria, along with those goose pimples, are the features of going "cold turkey". It is a dramatic, awful thing, and it's no wonder people go to such lengths, and depths, to "fix" their situation, and ride again and still this artificially created sense of happiness and euphoria. Un-beknownst, a road to hell.

On numerous occasions I have precipitated narcotic withdrawal in patients either by giving them certain narcotics that can cause this in addicted people, or as a cocktail to the unconscious patient who we later find has OD'd on the stuff. During their cold-turkey rant, it may be interesting to know that these poor sufferers also come out with some of the most vile, bitter, angry, and hateful outbursts that you could ever imagine. And many shit the bed. This says volumes about the contents of their personalities.

A Case History: The patient is a 36 year-old white male who is in the emergency department with the chief complaint of severe back pain. He has been hurting for about 8 months, and it all started with a fall at work. They told him that he had a back muscle strain, and prescribed a muscle relaxant and a pain pill. But when he attempted to go back to work a few weeks later, the pain returned and he has never gotten better, and hasn't worked since. He's had a couple of imaging studies of his back that have failed to reveal more than a "bulging disc", but it's his understanding it's these bulging discs that are hurting him.

For about the past 5 months he's been on pain pills that his family doctor is prescribing for him, but he ran out of them last night, and today his back is hurting him so bad that he can't walk. He thinks maybe there's something worse that's wrong, and his family doctor has "ordered him to come to the hospital".

He appears very uncomfortable, and soon asks for "something for this pain". He is fidgety and nervous, and looks like he's been crying. He relates his long sad story, and that he's spoken to his lawyer, and thinks he's getting jacked around. He claims his employer was somehow at fault (water on the floor, etc.), and he has contemplated permanent disability.

He's here with his fiancé, a decidedly low-budget-looking babe who tells me what a tough guy he is, and that she's never seen him like this before, and that something "must be wrong". When bedside examination indicates that the nervous system is in tact, and taking in the full assessment of the situation, the obvious diagnosis becomes narcotic withdrawal, and some sort of underlying behavioral disorder. Appropriately, I give him a hefty shot of dope, and soon he's back to the dude this girl thinks she's marrying.

After an hour or so, he's ready to go home, but requests a prescription of more pain medicine (vicodin, percocet) for this "flare-up". Now that I'm his friend, I confront him with his failed nurture, which is indeed the case. His dad was an alcoholic who was an abusive character, and he was exposed to violence against his mother. He wet the bed until he was 12. He's smoked since then, and his teeth are a mess. When I inform him that in fact these disasters create this form of dependent behavior and a chronic sense of suffering, he capitulates and says the same thing they all say: "Doc, all I know is I hurt". In this he has said it all.

If he persuades some unfortunate surgeon to give him "this damn surgery I know I need", he will find that the pain may be relieved for a few months, only to return "worse than ever" within half-a-year. But the studies show that if he'd just gone back to gainful employment, his social life would've stayed together, and the pain would've gone away like it does in normal people. But his absorption into negative thought and emotion, passive tendencies brought on by nurturing failure, and a no-doubt suspect support system has landed him in this stubborn self-defeating dead end we call The Chronic Benign Intractible Pain Syndrome. Though he believes otherwise, narcotics are a poison for him.

The level of dependency is astounding. There is dependency on the drugs, of course, as physical addiction. He's dependent on the medical system, and he couldn't care less about the bill he's running up, because he's sure not the one paying it. He's dependent on this poor lowly fiancé girl, for support and cheap love, and don't forget he'll need a ride home, after getting his "shot". And he's depending on his lawyer, and the Worker's Comp program, and eventually the federal disability agencies, for the greenbacks he'll need to eke out his marginal existence. And with any luck at all, he'll have to depend on street suppliers in the dark alleyways of nowhere, to stave off a stalking reality.

Case History: The patient is a 58 year-old white female who is grunting in pain such that all the patients and staff can hear her. It's her belly, and though she's been hurting for six months, it's getting worse, and it's never been as bad as it has for the past three days. Her doctor, who's been prescribing her pain pills, has "ordered" her to come to the hospital. It's her impression she needs to be admitted.

Again, that is. She's been in the hospital several times in this "horrible" past half-year, which has been "hell" for her. They thought it was her gall bladder, but weren't sure. But they took it out anyway, and though she felt better for a while, the pain never really went away. She wonders if her surgeon "knew what he was doing". Extensive tests since have been normal, but she thinks she needs another surgery, because that's what her family doctor thinks. (There is a myth that "scar band adhesions" inside the abdomen, ubiquitous after surgery, need to be "cut", and that the pain will go away then. The surgeons feel otherwise.)

I tell her that the pain pills her doctor is prescribing for her are powerful, addicting narcotics, and educate her regarding narcotic tolerance and withdrawal. She looks back at me in utter disbelief, instead insisting that "something must be wrong", and reminds me that her doctor sent her in, that he "knows her case". Though angry, when offered a "shot of something" for her pain and misery (narcotic tolerance/withdrawal), she responds "Please!"

When I have stepped out, she clutches her chest, and notifies her very concerned family members of these awful things I've said, that I've called her "a drug addict", that these problems were "all in her head". As they glare at me, and their mom lets it all hang out, I provide a description of The Chronic Benign Intractible Pain Syndrome, which basically tells her life story. They know that their mom has become a burden for them.

She reports that, yes, she grew up very poor, in a trailer. She never knew her dad, and her mom left her a lot to care for her little brother and little sister, since she was seven, and that she had to quit school as a result. Her first husband was an abusive drunk, but she "ran him out" and raised these kids alone. "I'm a fighter", she tells me. She promises she'll look up "this chronic benign intrack-whatever you're talking about", and I write it down carefully for her.

Diagnosis: Avoidant Personality. While some people set high goals knowing they will at least achieve something worth achieving, people with an avoidant personality problem will rather try nothing due to fear of failure or rejection. This is the same avoidance seen in phobias, which are a related disorder. And though they are people who want to interact, they are distressed to find they have a great deal of trouble forming productive relationships. This is unlike the schizoid, who is happy in his isolation, and different from the dependency of the borderliner, who will "go off" with rejection while an avoidant person will simply withdraw.

Why? Probably, these people have been undermined, or at least underinspired. But also, they must have trouble fundamentally somehow in perceiving others' actions and how their actions are interpreted by others. Their inappropriate quips will turn others cold to interacting with what are essentially the "dorks" of the world. You could call this behavior pattern the "poor social skill syndrome". Dumb socially, they are generally not dummies.

Diagnosis: Compulsive Personality. It should come as no surprise that there exists a personality where absolute control is attempted, where there is found a comfort in trying to control all variables of all situations. Such obsession is displayed in the human with compulsive personality disorder.

Unfortunately, attempts of this sort we know to be universally futile and self-defeating. Consequently, a chronic inflexibility leads to problems in relationships and in the workplace, and inherent difficulty in making decisions (with so much to take into consideration), and an inhibition of creativity. Mostly they are disappointed in the things they do, or may exaggerate their accomplishments or contributions to cover their feelings of inadequacy.

Most are conscientious and decent people, who achieve high in fields where these traits of responsibility and order are helpful. But most will find themselves uncomfortable adapting to change, or in situations where they must rely on others or where events are changing and unpredictable.

Why? Theoretically, when a child is made to deal with things at a young age, where he or she is unable to control certain events, though is made to, this results in efforts as an adult to demonstrate to the mind that it can control something. So relieved when it gets over this hump, it will now make a habit of it. Such chronic disappointment would be expected to bring about significant mood imbalances. This controlling, inflexible, inflated self-importance syndrome can be expected to be life-long.

OCD

In true OCD, there is ritualized, regimented behavior that a person performs in an effort to give the mind a sense of control over something, even though it might be over inane things. Cleaning and rearranging closets, repetitive hand washing, or light switch-checking, these all give the brain relief at a task clearly accomplished.

But alas, when there is significant intrusion into the social or personal realm, it is a distractive, distress-inducing malady. In its infinite variety and degree, there is some or a lot of it in our populace. For one person it can be incapacitating, for others an asset. Control, in all its forms, is sweetness to the mind.

There is also "thought OCD", where the conscious mind is continually invaded by threatening or anxiety-provoking notions, resulting in a spate of scenario play-outs and schemed reactions to events that never even happen. Perhaps there is no physical act that is done, and adrenergia is not playing a role. The mood level is high, and the thought generation is plentiful, but in a conflicted, contorted way. Learning may be inhibited none, a little, or a lot. It is a continuous scramble for sense among the chaos of an overactive mind.

Other Personality Types

Behavior Pattern: Passive-Aggressiveness. Persons with this maladaptation feign their ineptitude in an effort to carry out aggressive acts against others, and is a reflection of concealed anger and hostility. When they do agree to perform a task, they can be expected to undermine the project in some way. These people are essentially "negativistic".

Presumably an environment that combines violence and aggressive behavior with a "counter-inspiration" approach to rearing, such as intimidation by a parent, results in this stubborn, passive, and ultimately self-defeating behavioral disorder.

Schizophrenia

The term "schizo-phrenia" is a word that means that the mind is "split" from reality. It is a behavioral disorder at the far end of disorganized thought, a world of bizarre and disjointed fantasy, of totally incapacitating ill logic, of eventual long-term institutionalization. Please, don't ever confuse it with the completely unrelated condition of multiple personality (Dissociative Identity Disorder).

After an often uneventful childhood the young future schizophrenic will show the "prodromal" phase, where there is a gradual withdrawal from interpersonal interaction emerging over several years in the late teens or early twenties. During this phase there may be a report of "dysphoria" (anxiety, depression), and often illness behavior.

Then in early adulthood he/she will have their first major psychotic break ("episode"), and the diagnosis will be obvious. Most will declare themselves before they're 24.

Recall the description of schizophrenia given previously as the classic state of psychosis. The psychotic episode of the schizophrenic shows all the levels of disordered mentation: delusions (content of thought), hallucinations (perception), inappropriateness of affect, logic problems and loosening of associations (form of thought), preoccupation with an inner world of fantasy (autism), inability to establish goal-directed adult behavior (ambivalence), and abnormal posturing.

Usually some sort of life event will trigger the psychotic episode. Before modern anti-psychotic drugs these episodes could go on for months and a few years even, with eventual return to a more normal behavior pattern. But with each subsequent break, the schizophrenic will be expected to recover less well, and a step-wise deterioration will usually end up in total incapacitation and chronic care.

As many as half of all inpatients on the psych ward are decompensated schizophrenics. A classic experience is for them to do reasonably well while on their medication, only to stop it for some reason and slip into a psychotic episode within several months.

A problem in brain chemistry, schizophrenia appears to occur in all cultures at a fairly even rate. Most seem to be of the more paranoid variety, but they can range from silly (hibiphrenic) to flat out flat (catatonic). The chemical neurotransmitter out of balance is dopamine.

Kid Psychiatry

It is in fact quite challenging for the adult mind to see the world through the eyes of a child. Though all of us go through this maturation process, we aren't really all that sure what we see when we look back.

Think about it. What do you remember from your first year on the planet? How about two, three? Maybe you can remember a handful of things from first grade. Then more and more, perhaps. But even though there is so little memory from this early period in your life, probably you act like you act now because of these things you can't hope to remember, that happened before you started knowing how to remember.

A brief look at what's-learned-when goes something like this:

First year: Is the world good or bad? To get what I think I want, do I protest, or am I satiated and expectant (dependency issues)?

Toddlerhood: Can I play and explore, and learn (talk, walk)?

Four to six: Can I be social, express my wants and feeling in words?

Six to twelve: Identity development. Who am I, what am I. What will I do and what won't I do. Am I OK, cool even?

Adolescence: Who do I hang with? Do I trust my identity, or do I want to act more like someone else?

Adulthood: Goal direction. Social give-and-take.

It is evident that the earlier the error of the nurturing process, the more catastrophic the result, which of course makes so much sense to the Chaos-based philosophy, and the concept of fractalhood and the butterfly effect (Chapter 1). No one can say for sure, but we can either tell these traits early, or we've dropped some ball somewhere that so misdirects the fractal that the outcome is way off the mark. Personally, I think it's the latter.

Regardless of all that, here is a brief look at some of the more notorious adolescent behavioral problems. And it should be pointed out that many of these things seem to fade by the time the human is full-grown, at twenty, whereas many of the adult problems start around then, and seem to just get worse.

ADD/ADHD

The overriding feature of this group of wild-ones is the intense focus on the self. We say they're not "paying attention", but what we mean is that they're not paying attention to the things we want them to pay attention to. And the more we force them, the more evasive they get. What's wrong is in their difficulties being social, because they seem to be oblivious to the plight of others.

There seem to be three different kinds of ADHD, and this is my own artificial split of them. There are the anti-authoritarian kids they call "oppositional/defiant", pseudo-antisocials that have "conduct-disorder", and physically hyperactive poor-learners.

Case History: This boy is 16. He's got a little facial hair and is big and physically mature. He "went off" at school today, again, because the teacher told him to quit talking and being disruptive. He was sent to the principal's office where he threatened to blow up the building, and made a few suicide remarks, so he was sent to the ER.

When asked "what happened", he is evasive and has poor eye contact, and denies he said any of those things, and that he certainly isn't suicidal. The teacher's "a jerk", and his parents "are idiots". He thinks his dad's a pot-head.

His parents are both there, and they tell me that they've been having problems with the boy for years. He's always struggled to learn, and really doesn't like school at all. But for a few years he's had these outbursts when asked to do simple chores, and they seem to occur in waves, at some several-weeks pattern. He's had to be grounded several times for breaking various simple household rules, and the other night he took off in the car and only has his "temps". He's been put on a "mood stabilizer" by a child psychiatrist, and refuses to take it, because it makes him feel weird. He's been at times threatening, and lately he's been at his worst. They don't feel comfortable with him in their home and want him hospitalized, so he can "get help". He's ready to fight us rather than go there.

His mom takes anti-depressants, and his dad has been sober from alcohol for about ten years.

This child has "opposition/defiance" syndrome. There is an intense, reckless focus on serving the self, and the child has not yet realized the self-defeating nature of the behavior. He has responded to "control measures" by controlling, boring, dis-affected parents with defiance, and inspiration has been consistently lacking. He was a needy child in some key way, and the need was probably mismanaged. His parents have evidence of mild mental illness.

Case History: This little girl, 15, is brought in by her mother to get drug tested. Her mother knows she smokes pot, has been hanging with a boy who is 20, and she's failing in school. She "might be pregnant". Her mom is sure she's running with the wrong people.

Her language is foul, and she dresses slutty. Her facial expressions are that of anger and frustration. She embarrassingly scolds her mother in front of me, and tells her that she hates her. She has never worked, and her mother won't let her drive, which infuriates her. Her dad is "not in the picture", and has done time for "possession".

She has "conduct disorder". She likes the underworld of minor misbehavior, the thrill of the street, and the swagger of the bad boys. She is rather fearless, and has the type of relationship with her mother where they both "play each other", rather than share constructive thought.

Without their dad, girls can't know how to act. They lack the progression to more complex and sustaining behavior, a skill that will allow her to smartly avoid dangerous situations. What will eventually happen might include teen pregnancy, single parenthood, life on the dole, and lots of misery. And possibly, rape and murder.

Her mother may be ineffectual, but was nurturing in the key early years, so the child, while self-serving and reckless, is not anti-social. Yet. She's not "paying attention" to the signs of danger in her world, and lives for the moment, like her dad. They share a certain elevated mood level, and we recognize all of this as some sub-category of AD/HD.

Case History: This kid, man, he's too much, his parents are saying. He's like a ping-pong ball in a box car. The twos were "really terrible", and he was into everything. And since then, his decent parents have had to basically chase him his whole life. They couldn't keep him in several pre-schools because he won't leave the other kids alone. Now, at six, he seems to be having a lot of trouble learning in first grade.

This is the true "AD-HD" syndrome. The "stem" of the brain, just inside the bottom of the head, is where the "lights" of the higher centers are turned on. That big glob of gray matter you see when you look at a brain will only "come on" when it is stimulated from these "lower" regions. There appears to be a subset of kids where this area is underactive, and they behave as if they must literally move physically to keep themselves awake. Presumably, less of their brain can function in these circumstances, and they become socially retarded. The high energy, hands-on oblivion that follows looks to us like this condition of "AD/HD", but it is the true one. Stimulants in these kids are really beneficial, and can allow them to learn. As they get older, they seem to require the medication less and less, and then become mostly normal adults.

In opposition/defiance and conduct disorder, there is a parenting problem. In AD/HD, there is a chemical /neurobiological problem, and in all of them, there is a learning problem. Without abuse, the majority of these kids turn out OK, though minor mood disturbances and imbalance of personality traits can occur.

No doubt, young humans with this high amount of energy have the potential to perform at a very high level as adults. They are kids that probably need an entirely different approach to education, perhaps focusing learning on the things these kids want to learn, which is physical things, as opposed to the three R's. More and more, pediatricians are shying away from giving every kid that mom brings in one of the stimulants on the market that "slow them down". Almost every kid you give a stimulant (Ritalin, etc.) to will seem to settle down a little, and really, many of them seem to do "better", which is to do the things we want them to do. Why should we do this? It is surely a major intervention in the nurturing process. And you may be spoiling a little Einstein or Ruth.

Regarding Rhythm

Some kids will show off with misbehavior on such a rhythmic basis that certain researchers consider such syndromes some form of "bipolar disorder". But despite their unreasonableness, ADD/ opposition-defiance kids are not what you could call psychotic or deeply delusional, just immature and thinking too much about themselves rather than the goodies of competent social interaction. And their rhythm of mood sway is on the order of several weeks, while true bipolars only have a months-long episode a few times in their lives. So, if it is desired to recognize this rhythmicity in these not-yet-full-grown humans with their somewhat wayward pizzazz, maybe the term "bipolar disorder of adolescence" is appropriate, but in my opinion, is confusing and unnecessary.

Lastly, there are kids with deeper, unfortunate mental illness. From autistic children, to pervasive developmental disorder, and affective syndromes of childhood, there are rare and rather unusual conditions that offer a giant challenge to parents and the healthcare system. Most often, there is no sin of mom nor pop, except a wrong mix of genes.

Section 3:
Functioning, Of Humans

"Identity"

Perhaps the whole point of the chapter on Chaos is that individual units "organize" together as another "unit", in escalating size and complexity. At its scale, these units (fractals) each have a finite lifespan, shaped by the forces peculiar to its level of organization, during which time they exhibit and exude characteristics and qualities unique to them, and this is identity.

Often, looking back over the history of each such unit will reveal a certain "character", where it seemed to have things it liked to do, and things it rather didn't. This is true of objects and things, like cars and ballclubs, and of course it's true of each and all of us. Thus, when it occurs that we wish to ask, Why it do that?, we're asking about this thing we call "identity".

Feel free to ignore the omen if you dare, but early-on in the life of a thing you can already seem to tell a lot about it. We, as humans, recognize that longer-ago experiences and memories are particularly influential in the shaping of our minds, and our identities. The personality that results is defined by such "inputs" as respect (hero recognition, emulation), obedience (humility, insight), protection and nurture (clan pride), and desire for love and happiness. Once mature, we draw on the judgment acquired by the learning and evaluation of the results of one's actions on its environment that are serving as a groundwork for responding to new events as they unfold. Simply, they are memories.

Dissociation

Earlier in this guide to the creature we discussed the phenomenon of dissociation, whereby a person alters their identity through a trick of memory integration, in an effort by the mind to evade a painful reality. In these cases, where such a major deviation from a more stable selfness is playing out, the resultant behavioral phenomena are dramatic and cinematic, and suggest a number of interesting, far-reaching features of the wild ape-mind of humans. We explore this now in more detail.

A case history: I'm seeing a 14-year-old girl who's in the ER because she's "out-of-control" according to her mother. She's staying out all hours, is not responding to disciplinary measures, she's sexually promiscuous, and mom thinks she's doing drugs. When confronted with these allegations, the child says yes, mostly it's true, though she doesn't really like drugs at all, because they make her feel "weird". But she "might be pregnant", and her "guy" is "old, like probably 30". She's dressed in a sleazy outfit, her make-up is too much, and even with me she's flirty. She says she's "probably depressed", because, of course, that's what they've been telling her. They've put her on a mild anti-depressant, but it hasn't "helped".

So I ask her what is, to me, an obvious question: What happened? I ask her if she was abused in someway, and especially, tell me, were you sexually abused by anyone? She says she "thinks so", but that she has "flashbacks" about it, and in fact had no memory of it for years and years. It apparently had gone on from the time she was about 8 or so until she was 12, and it happened on regular occasion. It was her stepdad.

This is the classic "Dissociative Amnesia" patient. Unable to comprehend the fantasia of sexuality at this immature stage, the brain simply created an all new and different identity and blocked out entirely these painful and confusing memories that were experienced and "saved" by this "old" identity. We know all-too-well that such sexual exposure in childhood (pre-pubertal) is catastrophic to the nurturing process and the development of a strong and credible identity, forecasting a life of generally severe and chronic behavioral problems. Among them is dissociation.

In starting so much "from scratch", this new "means" of memory integration and storage, i.e., identity, will show poor judgment and insight into social performance, often taking on bizarre, outlandish, and dramatic character. The shallowness of it is apparent to all, including that 30 year-old.

The tendency toward flamboyance and behaviors traditionally recognized as decadent or taboo would indicate that "lower", more primitive and self-serving areas of gray matter are "having their say", this reckless behavior we are seeing. Why is it that these "id" centers are able to execute theirself where nurturing disaster has delivered a giant mental hurt? The mind, in its sense of bitter struggle, will be happy to explore such co-existent thoughts as exploitation and pleasure-grabbing. Why wouldn't it? How couldn't it?

Physically abused kids, kids exposed to parental violence, and emotionally traumatized kids are famous for bad behavior. But they don't have this amnestic syndrome. Why is that? Presumably it is the fact that sexuality and all its sensations and privacy are more consequential events for the brain to be fed, and the memory network they generate is deep and far-reaching.

Among the great observers of this phenomenon: Dr. Freud.

Another case history: The patient is a 35-year-old white female who is in the emergency department, brought by ambulance from the home of her sister whom she's been living with for several months. When asked to sign her check-in forms she is hesitant, and signs an illegible name. But she didn't know her birth date, and had to look at her driver's license when asked.

She is "not sure" why she's here, and her sister, who didn't come with her, spoke to the nurse by phone and said she has been acting very bizarrely for the past few days, and thinks she may be on drugs, but this does not appear to be the case. She said her sister, the patient, has been staying with her because she "didn't have anywhere else to go", since she and her husband broke up a few months back.

Then her ex-husband called. He'd met the girl a few years ago at a crisis center where he'd worked, and found an interest and sympathy for her even though he was aware of a great struggle her life had been. But she would fly into rages and act so profoundly abnormal at times that he finally concluded it was a gigantic mistake on his part and has been trying to escape her ever since. He's changed towns and required two restraining orders.

And he's calling us to report that he's done some research and that he's "certain" she has multiple personality. He tells us that her father had sexually abused her for many years in late childhood, has been some form of "lost" her whole life, and suggest we hospitalize her as a psychiatric emergency, which of course we did.

As she walked the halls to the bathroom a few times she would glance and glare in a spooky sort of way, hungry for any eye-contact she could find. She was not sad or angry, or anxious or depressed, just quietly and severely mentally ill. And socially incapacitated.

"Multiple personality", is now referred to as "dissociative identity disorder". The brain has managed to create a number of separate "me's", with separate memory management, with each of these identities storing memory under the bias of a delusional fact-bending, self-serving, maladaptive mind. And while many of them are oblivious to the existence of the others, some will actually communicate. They are generally archetypal characters, from ids and super egoes, and the way they serve the creature should be glaringly obvious. The lack of depth of these identities, and the inherent effect on judgment and action, make such individuals a concern to society.

Recall that "personality" is one's beliefs, insights, motivations, and loyalties. The longer this "identity" exists, the more depth will be provided by more and more memories. So as we get older, our identity gets more and more established and knowable.

But apparently, normal and singular identity development can only occur if these are integratable decent memories. If instead these memories are painful and bring shame, then a person's "ism's" can become distorted, and in some people, can be abandoned altogether.

New, shallow, and multiple identity formations show themselves as the concoction of some of these old isms, and an entirely new name and behavior pattern. Acquaintances, especially "loved ones" will begin to wonder if something's up when they see behavior that is so variant and different from that person they thought they knew. The cause in nearly all cases is sexual abuse during the nurturing process, a perpetration you now know to be catastrophic to the mind's development.

Fugue States

When I was a medical student there was a terrible local story about a petite/darling 25 year-old girl who worked as a reporter for a local TV news station who up and disappeared. When neither she nor her body was found for months and then years, abduction and murder was the assumption.

Eight years later somebody who knew her spotted her in an Alaskan airport, and when they confronted her with who she was, she at first denied it but then seemed to remember when "reminded". She'd married and had children, and changed her name. Her memory of her previous experience was very vague. She stayed in Alaska.

There was another story like that, of a middle aged man who'd mysteriously disappeared and turned up a month later, in Vegas of all places. Again, family had insisted that such an abrupt split was "impossible" from this person they knew so well. But again, Poof! Gone. New city. New life. New identity.

This is the "fugue state". Dissociative fugue is a condition where on one or more occasions a person suddenly "dissociates" from his or her past and its associated identity, and abruptly travels away from home. Though family and friends will find it impossible that this person is capable of such a thing, the behavior serves a purpose.

Fugues are "malingering" states, where a person does something he wishes to do, but "as" another person (identity). Like going to

Vegas, or Alaska. Or even to cover underlying impulses that could be embarrassing or dangerous, the identity is simply dropped for a new one. Whether it's a CEO disappearing to a dude ranch to escape overwhelming responsibility, or a common Joe looking to shed his commonness, the mind truly tricks itself and....just goes and does what it's got to go do.

Eventually, usually, a confusion will arise in somebody who's "fugued out". The original identity will return to the conscious memory, and notice a "gap" in time, and will be bothered by the fact that it happened, in some cases responding with adrenergia bad enough to present to an emergency department with symptoms. They return to their previous life.

A failure in the nurturing process will usually be uncovered in patients with this form of dissociative disorder, and presumably this mind would respond positively if it could explore the depths of these distressing and imbalancing stored memories. From a comfortable couch, perhaps.

With serious prodding, be it a patient in something like dissociative amnesia, or a fugue, or even a multiple personality performance, they'll all be able to be brought back to that identity that they started with and know so well. For many if not all, something like "psycho-analysis" is the only hope for lasting recovery of more normal behavior. But this fascinating process takes years and costs thousands. For most, a perilously shallow identity situation will mandate underachievement, social misfiture, and bouts with depression and chronic pain.

Memory, and Identity

People with dissociative phenomena show an apparent ease of hypnosis. And hypnosis, whatever it is, is marked by amnesia for the experience. Probably, this all argues for a more secular situation in the brain's frontal lobe regarding elements of identity and personality. There must be areas where we etch critical events that taught us defining lessons that we adhere to greatly, and probably instinctively, and desperately. It's similar to what would make you think about survival-type instincts, like keeping your hand out of the fire once you've burned yourself. Or avoiding dangers of the social realm. This type of memory might be called "long term" memory.

And, maybe, in another spot there's some chemistry for the moment, what we call immediate memory. There is an event fed to the brain, something seen, felt, heard, whatever. Are these familiar stimuli? How does past memory storage, located elsewhere, "interpret" this event, so as to store and learn relatively little or lots from it? Is there "insight", and is that somehow a connector for the two? How, then, does this "body" react, by what it says, or, maybe, does, to exact a control or engagement or, possibly, an escape of some kind from this "event" just now fed.

And there's short-term memory. Things we need to know now, but not things we need to know forever. Deadlines. Places to be. People to coordinate behavior with. In short-term memory, some brains will find some things much more worth remembering, depending on its relative personal value. And this will affect how this person behaves, thereby affecting everyone they know. "Oh, my bad. What's up? Was I supposed to be there? I, I forgot…"

Identity, then, is the consequence of interaction from our various memory banks, from the more essential primitive and protective centers, where a convincing event succeeded as a memory that inspired guidance, and to the more fickle meanderings of immediate and short-term "social memory" areas. It is more from these social areas that we do the day-to-day behaviors that our friends and co-workers see, with the resultant interactions what we call "personality". In dissociation we see them unravel, and separate from such basic memories as name and address. And past.

Regarding the mind and how identity and memory affect behavior in all social interactions, consider drugs with "hypnotic" effects, drugs which happen to be coming out of our ears in the form of alcohol, benzodiazepines (Valium and Xanax), and barbiturates, (the date-rape drug). They affect behavior when they are "utilized", and they do it by affecting identity, and these different sets of memories.

But especially, booze, a great name for our favorite troublesome beverage, affects our identity. That's why we like it. All of us agree that this particular substance, so finely titratable, will consistently "loosen up" elements of our behavior, allowing us to more easily say the things we want to say, do the things we just kinda wanna do, and, in fact, be the person we'd sorta like to just go ahead and be, until the shit wears off and we remember why, in a society, you can't do this stuff.

Perhaps you've been a witness to the folks who completely dissociate when they get boozed up. The identity now in control of the consciousness might just go violent, lovey-dovey, mouthy, edgy, boisterous, whatever. Later, when it sobers, and no memory remains for the behavior, we say a "blackout" has occurred. Why such ease of dissociation under the hypnotic effects of alcohol?

Since alcohol, such a little chemical, presumably reaches all the brain's tissues in a similar concentration, perhaps the area for identity and memory integration is less-represented in these people, or maybe such tissue is just more sensitive to the stuff. Whatever the case, it is alcohol's effect on memory, and therefore identity, that accounts for why this substance affects people so differently.

Alcoholism

One more important observation regarding identity: alcoholic families. If dear old dad is a drunken shame and a source of steady negative stimuli, juniors in the family will have horrendous problems with identity. And dad, he's obviously got his problems with identity, and if there's a genetic basis for alcoholism, it might involve the chemistry of identity.

With alcohol, he can effectively escape the problematic and painful memory integration of his current identity. Perhaps he is escaping boredom and a chronic sense of hopelessness, or of "dysphoria", or the good old fashioned term, "melancholia". Whatever the matter, he finds himself easily hypnotized by the effects of this little chemical, and the family notices how it strips him of his connection with some of the more important necessities regarding his immediate and short-term memory pathways. If it wasn't for that, he wouldn't do the things he does, and say the things he says, which are undermining and confusing and detrimental to the formation of strong, socially-acceptable identities that bring people productive and satisfying social interaction.

"Drug Use and Abuse"

Humans, like other animals and plants, naturally want to do what they want to do. In forming America, our forefathers acknowledged this and allowed for the "pursuit of happiness" as a right "inalienable". The "freedom", to do as thy please with thy damn self.

So, they partake. "Substances" have been a part of the plight of humans on Earth forever, and will be forever. It ain't like you can stop them. They'll eat, smoke, swig, huff, puff, sniff, shoot, pop, or skin-pop even, any means necessary, headed for wherever the hell, and why-ever-the-hell. Within reason, and by responsible folk, substance utilization is fun, and that's why we do it.

The Modern Age, in all its fanciness, will offer mankind more and more "modes of escape", and probably modalities of escape as well. Indeed, his "unnatural environment", and all its torture of thought, also requires the need for escape, providing motherhood to both the necessity and the invention.

Tempted endlessly, hounded continually, see them duck, and slide, and groove. And then rally. Most humans maintain functionality and a balance of play and productivity, and achieve a sense of fulfillment in the face of such inherent dangers. What they all need, mainly, to pull it off, is a sound nurturing, raising again this word "mother".

But plain and simply, some people have a little trouble getting the dose right. Or the dosing interval. They may hurt themselves in some way with a chemical, or get under its influence, lose their social skills, and violate somebody's civil rights, or break the law. Or miss work. There comes this time, when it's official: there's a problem here.

People reveal to health care professionals both the embarrassment of the extent of their struggles with a defeating addiction, and the psychodynamics of their failures as well. Such confidentiality allows them to tell us the truth, best they can, about how a "substance" has now become a medical condition for them. Many express that they're ready to be a new person, one proud, mainly, to never be that old one again. It is hopefully a situation of great enlightenment for them, a time when they can finally accept that this substance is a reflection of a behavioral condition, an error of personality and thought. A something missing.

For so many, it doesn't happen for them. They do OK for awhile, but before long their mental problem manifests again in some way, with yet more forms of bad judgment and miss-step. For some, the bottom comes in the firmness of a hard box.

While societies do their best to reduce the dysfunction and self-defeating behavior of the strugglers and stragglers, the history of such efforts, in this country at least, have consistently made matters worse. Attempts to "abolish" the utilization of alcohol created a mob we would never get rid of. The "war on drugs" of a later generation made criminals of the innocent, many of them mentally ill. In treating one social problem with another, we get what we basically begged for: another mob.

Destructive public programs like welfare and poverty-concentration have created factories of nurturing catastrophe, resulting in a population so maladapted that resistance to substance utilization and association is almost impossible, with many now never a witness to "normal" behavior. "Gangs" fight over "turf" for their substances, creating a no-man's land in the heart of civilization. A ghastly hole of negative energy, vast sins of the local moms and pops. We blame the substance.

But you knew all that. What follows is a little insight into what it's like on the other side of the rail of patients we see when they are having to access the medical system because of "miss-usage" of some "substance".

Tobacco Abuse

Back when they let patients smoke on a psychiatry ward, upon entrance you'd be met with a wall of cigarette exhaust that could part your hair. Know this: all psychiatry patients smoke. Make whatever assumptions you want, but they all smoke.

Nicotine is the non-specific tonic for the troubled mind. In social situations, when you're nervous, or when the conversation deepens, about a third of us will fill the moment with a dose of nicotine. And.. Ahhh, they feel better, for sure. It's a drug that works.

Addiction to the calming effect of nicotine is legendary. We know that, chemically, it is profoundly addicting. Once ritualized, it is probably the all-time hardest habit to break. If "they" hook you early, you just might smoke a zillion of them over a lifetime.

Ever find a cigarette butt in your potted plant? And a smoker might Oops! litter the spent pack, which they have now become so careless with after so many cartons of them. And they'll happily blow smoke in your face at the ballgame, too. The fact that they appear oblivious to the plight of the people who suffer from their local pollution is another example of a focus on the self, and the moment, so typical of so many mild and major behavioral disorders.

Is nicotine somehow making them want to focus their behavior on themselves? Is it a tonic? Or is it a tonic that creates the need for more tonic? If every schizophrenic smokes, will we ever know if they'd be one had they not smoked?

If cigarette makers have enriched their "product" by boosting this chemical, with its heavily addictive quality, then they have committed a large crime against their society.

All of this, and we haven't even talked about the awful effects it has on those who utilize nicotine's delivery system, the smoke. The gas, that is, and the unit itself, which "the user" refers to it so warmly as. (Ironically, his fellow utilizers generally don't want to hear the old "Hey man, can I bum a ...smoke?")

The dirty burnin' is consistently manifest in the non-sparkle of that key asset: the smile. And patients who come to their ER for toothache have this in common: they have limited financial resources, a history of underachievement or self defeat, and they want narcotics. And they all smoke. A dirty mug belching dysphoria.

Mouth and throat cancers are usually diseases of smokers, and especially of smokers/drinkers. The same is true for Ca of the stomach and of the colon. Lung cancer is almost unheard-of in nonsmokers, despite what you hear.

The effects on the circulatory system are possibly even more pronounced. Heart disease, especially early heart disease, is very often from a smoking career, and "events" like heart attacks are precipitated by this habit. We note that people who exercise and pump up this vital system and fill the bod with oxygenation get much less of these problems, while smokers are very rarely in good physical condition.

In the lungs, a state of chronic inflammation exists for smokers. Bouts with wheezing and bronchitis are much more common in cigaretted lungs, and the progressive scarring caused by this chronic inflammation results in the much dreaded misery of emphysema, the end-stage lung. Can't-use-it tissue. With every single breath an adventure, it is a plight that mimics the smoker's moment-to-moment existence.

And for what? A million tranquilizing moments, presumably. The "stuff" of hard-core addiction.

Perhaps. But a smoker's main addiction is to taking breaks.

Alcohol Abuse

Now, alcohol. Or "alki-haw". It is one of the great ironies of American Society that the two most destructive of "substances" are the two legal ones. This second one, ethanol, is a monster among us. For those with the "ism", a trail of tears and a huge drain on society. For the rest of us, the stuff only costs us a major disaster a time or three throughout our lives.

Ethanol is a small chemical. Its infinite solubility in water allows it to penetrate all the watery systems of the very watery human, where it consistently is toxic. It can best be considered a pickling agent.

Humans like it for what it does acutely to the "mind" part of the brain, where it is a depressant. The resulting release from the inhibition of learned, taming instincts (logic, judgment) allows for a brain more pliable for play, and more "lubed" in the social situation. In other words, drunkenness.

Over all, the toll on us as a result of our utilization of alcohol is large. Not only is it a major factor in serious auto accidents, domestic violence, bar-room brawlhood, and adult swimming fatalities, but also the price paid in complications from liver disease and pancreatitis, as well as other degenerative alcohol-induced illness is big as well.

But the largest cost to society, by far, is all the dysfunctional children of alcoholic parents (fathers especially) who typically grow up to have poor identity and judgment, failed relationships, and chronic bouts with anxiety and depression. Alcoholism in them and in their mates is terribly high as well. Underachievement, self-defeating behavior, and limited resource are characteristic baggages for adult children of alcoholic parents.

The long-term effects on the nervous system include diminished sensation, loss of coordination, a wide-based gait, and eventually too much gray matter loss to allow for enough remaining IQ points to craft a sustained run of sobriety. Hence, for many its a ride to the grave.

I used to think that alcoholics were sad, depressed people. But it is increasingly obvious to me that there is a "hyperness" to most of them. They have minds that generate a lot of thought, and they show impulsivity and poor judgment. Many have some form of "ADHD"-type behavioral problem, that intense focus on the self. I wonder about the role of such an active brain, and figure that it is their inability to master this activity that defeats them. You could argue that the "social IQ", the capacity to learn how to act, is questionable.

Chronic depression, repeated DUI offenses, job-performance interference, sexual dysfunction, increased incidence of malignancies of all kinds, and a shortened life-span are typical co-morbidities for alcoholics. And, of course, they all smoke.

Alcohol withdraw is fatal in 20% of cases if untreated. In binge drinkers (heavy liquor in large volumes, daily, for weeks or months) there is such depression of the nervous system by this agent that when the alcohol is withdrawn, the nervous system absolutely "goes off". After a day or so, its the shakes (tremors). The heart begins to race, the blood pressure rises. Even fever can develop in this hypermetabolic state. By day two of this even a full-blown seizure can occur (the "rum fit").

By the third day there are often dramatic visual hallucinations (fire-breathing dragons are much more common than pink elephants), delirium (confusion, disorientation), profound tremulousness, and autonomic hyperactivity that drives vital signs to dangerous levels. If the heart and muscular arteries cannot maintain this output to support the demand (and drunks are always malnourished), then circulatory collapse brings about a fatal situation.

In the old days they put people on alcohol drips and backed them down slowly, and this works. We use sedatives from the valium class, and this also is effective. Still, you can die of this alcohol withdrawal syndrome (delirium tremens, or "DT's") in the best of hands.

If you ever find yourself on a serious drunk you may notice that your nervous system will rush a little as it wears off along about 5 a.m., and you may also find this "agitated rebound" can be calmed by a little sedation, like a beer or a drink. This is where the saying "hair of the dog that bit you" comes from.

The hangover is a mysterious ailment as far as explanations, so here's another: it is this agitation and nervous system rebounding that results in very limited restorative sleep, and hence the next day's major fatigue, headache (extensor objection), and nausea (pickling). It takes about a half of a day to recover.

Elixir? Salicylates (esp. aspirin) before bed, and maybe a caffeine source. During the agitated rebound, a stretch and a walk, along with a carb snack. And ideally, a sneeze or two.

Prescription Drug Abuse

Third in the order of trouble: fancy chemicals like pain-killers and tranquilizers, available by trips to the medical system with just the right manipulative powers and outright lies, to just the right doctors. When word gets out that somebody is willing to prescribe them, every crock in the city is on the sign-in sheet within weeks.

For most prescription drug abusers, it is a life of failed nurture, personality disorder, sick-role playing, and illness behavior. Passive-dependency, thymic depression, and anti-social personality are the typical clients. Job-related or tort-inspired injury are classic signals that another of these creeps is here to shamelessly take what they can take.

ER and doctor-office shopping can net enough pills to run a small business on the street where there are a plethora of dysphoric, dysfunctional, self-destructive mouths to "feed". Back pain, shoulder and neck pain, and headaches are the preferred malady of these less-than-forthright exploitative, negative energy characters. We suffer them.

When not outright selling and abusing them, certain subsets of abusers will "utilize" these thought- and sensation-altering pharmaceuticals enough to essentially re-tool their minds altogether, and while many find temporary relief from the voids and depths of mental sickness, most if not all will find the slope is slippery, that falls are the norm, and that a new set of problems will now exist.

Unfortunately many are "sanctioned" by their doctors as "legit", ostensibly based on an X-ray or scan finding, and agree to make narcotic addicts out of these folks who have become a regular customer at their doctor's private medical business. As tolerance develops, the misery of the up and down dosing begins.

Let me be real clear here: if you're dying in a few weeks or a month, we ought to dope the absolute hell out of you. But if you're on daily narcotics for any other reason, you've got a serious and difficult chronic mental disorder. Period.

The Methadone Phenomenon

Methadone is a narcotic. It alleviates pain and provides a sense of euphoria. It is addictive. In specialized clinics, chronic narcotic addicts can enroll in programs where they are basically controlled with smallish, steady amounts of this drug. It's controversial.

I remember this guy one night. The face of the sociopath, the tattoos, the body. He's done something minor, and he's in the ER, and I ask him, tell me about this methadone you take. He says, "Well, it keeps me outta the joint." Laughs. "No really, when I take it I don't have to get high on heroin. I can work, be a dad". I found it interesting.

Generally, there is a history of disasters in the nurturing process in people like this. Divorce, alcoholism in the parents, neglect, and abuse are typically the ingredients of adults with chronic dysphoria, where escape through substance use and abuse are understandable alternatives. The exploitative nature of their gig, and their frequent-flyer status in the medical system are in line with such mental illness.

Cocaine

In fact, there are two separate phenomena that surround the use of this chemical in American society. One is the powdered version that is snorted, and the "other" form that is smoked in a pipe. The powdered "snow" is preferred by the more sophisticated, higher-income users, while the smoked "crack" or "rock" preparation is more a poor-man's thing. In either case, an instant "buzz" results, and a sense of euphoria exists for something like 20 minutes to an hour. Casual use by either is generally safe, although with cocaine there's always that worry, that rare person who gets funny heart electricity and drops over dead after doing it. So, you know, good luck.

For many, a horrid addiction syndrome occurs. If we're playing stereotypes, which is easy and fun to do, they go like this: "Snowstorm" syndrome is a condition affecting middle-aged white men who fancy the "high" life beyond their ability to keep it all above board. Before long his finances are wiped out, his family is wiped out, and some important neurotransmitter pathway in his brain will never be the same. Actually, he will recover pretty well, and unlike the typical hard core drunk, coke-heads generally learn something they needed to know, in a very hard way and live to tell about it. A majority don't do it again.

Crack is the poor-man's coke buzz, and as this inhaled form it shows an absolutely wicked chemical addiction/crave syndrome. Inner-city police will tell you stories of how low a person will go to get that next rush of euphoria. The nervous system injury is worse from this mode of delivery. And more commonly than with the powdered stuff, women of poverty are crack-susceptible.

As opposed to the kinder, gentler surrounding of the typical snowstorm syndrome, the immediate environment of the average crack house is a war zone. Generations of nurturing failure have manufactured a sub-culture surrounding the rock-of-a-drug that has resulted in lawlessness, hopelessness, and senselessness, and a despair so deep that few can grasp. You could call it illogical.

Judgment: devil drug. Avoid at all cost.

Marijuana

From a medical standpoint, it's no secret that this illegal substance is a harmless pharmaceutical. The "cannabinoids", chemicals in the weed, are hallucinogenic in the human brain. As a thought stimulator it treats its utilizers to a flood of sensory augmentation, enhancing tastes, smells, and sounds, as well as causing "flight" of ideas, and a rambling, often loosely associated train of thought.

Minors who smoke a lot of dope are making a mistake. For many, school performance can be expected to drop off, and important teen-parent interactions impaired by the divisive nature of the catch-a-buzz routine.

Police officers will tell you they don't see pot-heads causing a lot of "domestic disturbances", owing to the mesmerizing, tranquilizing effect of the cannabinoids. In a generation or so it won't be a crime anymore to "do" the stuff, but surely the loss of such social naughtiness will take some of the "fun" out of it.

"Junkies"

In the old days, heroin was called "junk". As an IV experience, this intense, wild run of gigantic narcotic highs, of orgasmic euphorias, draws in its victims despite the very large dangers of the pastime. Besides having to thieve and beg for the large dollars necessary for the habit, they face the dangers of catching hepatitis and "the AIDS". If the dose is too high, that narcotic might just make you lay down and forget to breathe, and the lights go out on your nutso party.

But IV drug users are, themselves, their own set of people. Often intelligent and "far out", many appear to be saddled with a major dose of dysphoria, and they find this drug-induced reversal irresistible.

AIDS, Hepatitis B, and Hepatitis C are common in IV drug users. They also get vein clots, pus at the injection sites, and even infections on heart valves. They are generally frail and chronically-ill-appearing. They represent a very dangerous part of our job, to care for them and be near their "system" that is escorting around these world-class germs.

Speeders, Trippers, and Designers

The "wilder" set of relatively harmless chemicals, these profound brain stimulators are only a mild nuisance so far as medical system visits are concerned. On occasion we'll see some "really flipped-out dude" who's gone into either a rage or some dysphoric experience like extreme paranoia. With a little dose of anti-psychotic medication we can…take the wrinkles out of their face until their metabolism has removed the substance from the circulation. Most of the time these kids settle down and go home.

While rare utilization of this modality of brain entertainment, especially for well-adjusted, well-nurtured, mentally sturdy stock appears harmless, repeated and frequent use is likely to cause a mental injury, with paranoid, anxious, and depressive syndromes that can be chronic and recurrent. This risk of "losing some of your mind" to "trips" into psychedelia should make young daredevils think twice about use of "acid", "meth", or the new troublemaker, "ecstasy". They can definitely hurt you.

Solutions

Simply, a society should put all its resources into any form of betterment of the nurturing process, and this is not only for the sake of decreasing substance abuse behavior, but for all the perks of a mentally- well population. Social programs that promote single motherhood should be discouraged at all cost. Young parents should be relieved of tax burden, and as a couple they should all be afforded "a home". Matrimony itself should be heavily scrutinized. Moving beyond public education to smaller, more spirited and practical curricula, with integration of the workplace, would make for more well-adjusted teens, and offer a better chance at rescuing straying souls.

Once broken, fixing the substance abuser is an adventure. When destructive and self-defeating behavior (the definition of substance abuse), requiring the services of law enforcement or the medical system, is centered around a substance (like in alcoholism, pain-pill popping, IV drug use, snowstorm syndrome, crackheadedness), often environmental remanipulation and psychiatric intervention can show a stabilization and return to productivity.

And most importantly, we must all never forget that yes, it does take a village to raise up a young human. An off-hand comment or compliment here or there could offer profound enlightenment to someone in this way needy, and we must all do our part. Because as the "Global Village", and all its wild wackiness, comes finally into focus in this New Age of the human condition, telling the difference between heroes and hacks, between truth and slick fibs, between positive and negative energy sources, will all become survival skills for those of us yet to come.

"Flexion v. Extension"

Recognize first, if you will, that the human is decidedly different on front versus back. Throughout time, the back of creatures has had to serve the function of protection, from the elements and predation, and in fact our own back reflects this ancestry. The big sturdy spine is there, the protective ribs and wide wings of the pelvis and shoulder blades are there, and the skin is harder and drier. Though not as dramatic as the stegosaurus, we too will benefit by turning our back to trouble.

This "dry" back, the anatomic word for it is "dorsal". On the other side, the "ventral" surface, we find not only a plumper, fatter skin, but also all the un-protected gut, the heart beating in the bottom of the rib box (just deep to the sternum), and the perilously vulnerable wind pipe in the ventral neck. The face is also ventral, and find there the sense organs as well as the intake for gas exchange and foodstuff. The nipples and reproductive system, as well as the umbilicus are also found here, where you would expect to find them, ventrally.

And this. The exit of the digestive system is conveniently, exactly, and just barely, dorsal.

Lastly, don't forget that the human, like most animals, had its legs flipped around somewhere along in the evolutionary experience, making their dorsal side face the front ("anteriorly"), and the juicy side the back ("posteriorly"). The soles, like the palms, are obviously ventral tissues. The nails, you note, are dorsal items. From foot bones, to the shin and kneecap, and even the femur, they (the bones) are just below the surface like you see as a dorsal phenomenon.

OK. Now know this: the muscles that occupy the dorsal side are markedly different from those that are ventral. And, as if to make it easy to understand, they differ in the same way that dorsal and ventral themselves differ. Introducing the flexors, which are bloody, juicy, powerful, and ventral. And to re-set these beasts of burden, the opposing extensors, which are broad, flat, thin, weak, pitiful, and, don't forget, dorsal.

Throughout nature we continually see this oppositry, coexisting in order to form a dynamic dichotomy. "Mates" working together to do what neither can do separately. We see it in flexion and extension. Musculature mates. They are quite opposite, and they each have their own characteristics, things they like and need. They are discrete units, flexion and extension, duking it out to do work and move and generate power for the human.

If you were to speak in terms of Chaos, you can see fractals, and recognize the levels of organization. At the atomic and molecular level, certain chemistries and energy supplies must be provided to allow for the biology of muscle activity. Each individual muscle has is own set of perils and job descriptions, i. e., its own plight and history. And there are muscle groups, occupying regions of the creature, and they are subsets of flexion and extension as a whole. They together embody the musculo-skeletal system, itself a subfractal of the human.

Caution: Muscles at Work

Muscles. We know they "flex", or "get short" as a great anatomist once observed. There is energy expended, and the structure does work. A joint moves, etc. A better word for this is "contraction". Because flexion has an entirely different meaning anatomically.

"Flexion" has a powerful motion that essentially closes to the ventral. Joint angles are flexed in acuteness with this bit of work, and it is the reset of these muscles and joints that is the work done by the extensors. Usually an extensor is opening up the human.

Over the years, the more hardy ventral structures, the flexors, will "hold up" better, while the dorsally-derived extensors will whither and atrophy. As the flexors overmood the extensors there results this slow but sure progression from a generally extended creature in youth to a flexed one at the other side of life.

Whatever the reason, the bod "falling into flexion" is the reality, and any number of famous ailments result. Remedies might require a certain philosophical discussion into the nature of things. The realities of aging, of dorsal and ventral, and of flexion and extension.

The Shoulder

A good illustration of the plight of flexion vs. extension takes place in the rotator cuff muscles of the shoulder. Originating from the scapula, the shoulder "blade" (dorsal), are several muscles that rotate the ball of the humerus. A few are flexors that rotate the joint internally (toward the soft underbelly), while several rotate the joint externally as true extensors. Holding the scapula to the spines of the backbones are another set of extensors that try to hold the shoulder blade in place during arm motion.

Ventrally, the flexors are engaged in activities that generate power, like trying to throw a little ball a hundred miles an hour. The pectorals, the biceps, the front of the deltoid. Umph! On the backside of the shoulder, the rotators, located inferiorly to where they insert in the top of the arm bone, pull the shoulder that direction, inferiorly, giving athletes the down-sloping shape.

As the extensor side loses mass from disuse and aging, the shoulder blade migrates superiorly, and the shape of an older person begins to appear. Eventually the top of the humerus pinches the rotator cuff against the collarbone, and now you can't throw like you used to. Or raise your arm like you did in class. It is…Oops! A little slip into flexion.

The joint is unstable anyway, held together only by muscle. If these muscles, ahem, haven't worked together recently, a lot of use will tear things up in there, cause inflammation and scarring, and a lot of pain with a lot of activities. With toning and strengthening, and good maintenance, the joint can last a lot longer.

Maybe the best exercise for this poor group of extensors is the pull up, and it's well known that the average middle-age human cannot do a single one.

Tennis Elbow

The condition of "tennis elbow" occurs when the extensors of the wrist and hand are pulling away from the origin at the elbow (distal humerus). Broad, flat dorsal, and pulling out of the bone. That's extension for you. There is no such ailment of the flexors there.

Virtually everyone who is at all active with their hands will experience bouts with this common ailment. Eventually the hurt makes you stop "overusing" these extensors, giving the microscopic tears time to heal. Until the next time.

The Knee

The leg muscles on the front of the upper leg are classic extensors. Remember, dorsal, even though anterior, since the human had its legs flipped around. They call this muscle group the "quads", since there are four of them. One is the sartorious, a long skinny muscle that lifts the leg to put it in the position a taylor needs to so he can work on clothing (hence the muscle's name). It is a move you don't see a lot of old people do.

Another of the quad is the bulge of muscle just above the knee toward the inside, the "dancer's" muscle. It is the "vastus medialis" muscle. There are two others, and as a group they perform like extensors, with their flexor counterparts being the classically ventral hamstrings. As you walk down steps these front muscles ease you down to the next step, and other important activities where these muscles "control the flexion" of the knee joint, in this case battling gravity.

Over time the weakening and atrophy of the extensors leads to a decidedly different gait. The shape is different. The stride is shortened, the dance step much less spectacular. As movement becomes more constricted the observer notes an "older" appearance.

Flexor-extensor imbalance in the thigh results in problems in the knee, since this is a joint held together as a hinge. Rotational injuries that tear soft tissue (notably a meniscus) become commonplace with activities that were previously much easier, a condition we refer to as "weekend warriorhood".

When the hip is flexed, propelling the leg forward, it will land you on the out-stretched foot, and hope to God your extensors are up to the task of "landing" you, by controlling flexion at the knee. At the end of their careers athletes have this difficulty landing and moving on to the next step.

It is worthy of note that the muscle doing this flexion at the hip is a big fat round juicy muscle called the psoas, and you may notice you can easily hold up a TV while opening the door for yourself. This powerful flexor is snipped out of cows as what they call "filet mignon". When we're eatin' good, it's a flexor we're having for dinner.

Calf Stories

The lower leg plays the dorsal-ventral thing as well. The calf is the big juicy flexor, with the extensor group on the front-to-outside of the leg. The calf muscles, arguably the strongest of all, eventually pound the extensors into oblivion. And by pulling on the heel bone they stretch the fascia on the bottom of the foot so hard as to result in the famous condition of "plantar fasciitis". The treatment: calf-stretching. Standing on the step. Stretching is always something good to do to a flexor.

When we reach oldness we end up doing most of the walking with our calf muscles, while the rest of our extension is so poor that if we get very far out of plumb at all, we fall. Falling is a major issue in care of the elderly human.

Miscellany

In cerebral palsy, the child loses higher nervous system performance such as the control of muscle tone. With the muscles free to "go spastic" the flexors predominate, and the result is flexion at the knee, ankle, hip, elbow, shoulder and hand, and face. Can you see the natural override of flexion?

"The clap" is a condition resulting from long-term syphilis infection, where a peripheral neuropathy is especially felt by the extensor. The noise is made as the un-opposed flexor smacks the foot against the floor.

Likewise, in the peculiar neuropathopsychiatric illness of multiple sclerosis, extensors appear to be more "affected" than flexors. When nerve supply to any region is hurt, the flexion will over-ride extension as the clinical finding.

The Back

There's something about the back. If you're buying at all into the different-since-dorsal observation, the spine itself is funnier yet. And located so centrally in dorsality, and in the very back of the front, and with so many bones and so many muscles, and all those nerves, much can and does go wrong.

In youth, a limber back accomplishes incredible amounts of complex movement, serving as the axis upon which the bilaterally symmetric human casts its flexion against its extension. In addition, it connects and coordinates upper and lower extremity activities, and can absorb shock with its intervertebral discs. In the neck there is 180 degrees of range of motion. And all this, wielding extension.

There are two masses of muscle on either side of the midline in the back, and they are composed of several long, spindly muscles, some spanning just a few levels, with others spanning 9 or 10. They help straighten the spine (extend the whole human), and control rotation of the trunk. Furthermore, they have a series of veins that flow by and pick up the heat they generate, dumping this hot blood into the thorax and the heart, and this keeps us 98.6. If needed, they will shiver when you go outside and get cold, or when you have a fever, to heat you up.

And don't underestimate the power and resolve of their biggest enemy: gravity. Ultimately, you know who ends up winning this titanic struggle. It is big work these muscles do for the human, and therefore it is probably not a surprise to know that perhaps the most common complaint we see as generalists is back pain. Usually, one of these muscles has torn something inside them, and the strand has gone into a spasm as a result. Intermittently, the thing will even contract full scream, creating what most people feel is the all-time hurt. After several seconds it will calm down to a less painful spasm. This is one angry extensor.

A typical pulled back muscle will hurt bad for 4 or 5 days, and then somewhat bad for weeks afterward. With the injury usually caused by a failed attempt to control a rotational duty with a limb extended, it is rotational stretching of the trunk that will pull this spastic strand of extension to where it can work again without sending pain signals to the brain.

Eventually for most of us, our back hurts a lot. It is the cry of the heart of extension, that it isn't totally handling what it is you're wanting it to handle. And it is usually suffering from too much of the load being heaped upon it because of over-extended extremity extensors, and over-ambitious flexors. And perhaps some degree of inactivity, or over-activity.

Chronic Back Pain Syndrome

It turns out there is an epidemic of disabling work-related back injuries, and it is especially curious that these "syndromes" are seen so much in people with a history of poor nurture, passive-dependent personality styles, and symptoms of depressive illness. (Society "enables" the disability phenomenon, hence the epidemic). Can the plight of this great extensor be viewed as a metaphor for exhibitions of flexion vs. extension? Look hard.

Again, flexors are strong and businesslike, while extensors are broad, flat, and poor, and require special attention. Consider flexor stretching, and extensor strengthening. "Use" your flexors, and "manage" your extensors.

Headache: Extensor Objection

There's an old saying: "If it's my headache, then it's a migraine headache." Migraine syndrome, a blood vessel phenomenon, is fairly easy to diagnose. But many patients who say they suffer migraines are being pestered by the musculoskeletal phenomenon we call "tension headache". By this we mean that "tension" is the act of these muscles, in a tightened state for some reason, pulling on the head muscles in a painful fatigue. And while various concoctions have been created for relief of this extremely common complaint, what is happening is that the body is calling out for rest. It is the only real treatment for the condition.

As an expert on the syndrome of skeletal headache I have uncovered these headaches for what they are. What is happening is a battle of flexion and extension, the latter no longer willing to work. The source of the hurt, of course, is coming from the extensors, as a tensed-up pull-ache source of pain. They are insistent, expectant, and will pull the creature into a down position, hoping that with enough sensory deprivation that big fat glob in the head will go into the "off" mode, and sleep awhile maybe, thus allowing the skeletal system to rejuvenate itself.

Up the back, to the lats and the traps, and the shoulder extensors, and that part the deltoid that's back there, to the neck, to the head muscles, it is a great trail of tears. Besides tiredness of spine, be careful not to underrate upper girdle over-use and its contribution to the occasional mass-mutiny of extensor objection and skeletal headache.

People who fight mental problems like personality disorder and depression, and anxiety, all have a hell of a time resting, and many of them suffer headaches. Since they've spent a lifetime living "an act", they continue these poor coping mechanisms and illness behavior, and treatment with medications that make the problem worse, like tranquilizers and pain pills. All the while they're blaming "those migraines".

These people are so far out of the groove that they can't rest, suffer fatigue, and are hounded by their extensors who are demanding such special treatment as real physical activity and conditioning. These muscles may even be well served by such lavishments as massage, chiropractic, and oriental babes doing a walk on them.

Reducing stress helps. When the adrenaline tone is high on the bod, the consequence of increased alertness will result in what is sometimes a wicked run of skeletal headache in even normal people.

Alcohol, an awful drug, is also trouble for extension. When the depressant wears off, along about 5 am, there is a rebound alertness and wakefulness. The resultant headache is an extensor objection. No rest, no play.

A "mind" (brain) that is unable to find a rhythmic and healthy flow of thought will be unable to distance itself from the alertments enough to do what it must do, and that is rest. It is that parallel never-never land where all of us must go, and can't do without. If we don't, we suffer extensor objection. (Of course flexors need rest too, but they don't cause headache.)

Don't forget, a horse will stand there on all fours and sleep 30 minutes and is ready to pull a wagon across the prairie. Dumb animal, we say.

Going Further With Flexion and Extension

We see, then, that before our eyes we have illustrated for us that something the universe does so well and often: dichotomy. On the one side dorsality, the other ventrality. Both perform as a plight, with their own set of circumstances inextricably tied to that of the other. In this case, flexion vs. extension. For sure, a balance. We look now further into the dichotomy, with a ramble on f vs. e.

In the face, smiles are extension, while flexors are happy to frown for you. See the over-worked flexors in all those folks out there fighting the battles of nerves. With exercise, the smilers can be expected to tone up. (And don't forget gravity wants to pull the face down too.)

In chronic sadness, the flexors are so overdriven that extensors seem to whither away to nothing, and the face itself looks different, and in a way we all recognize the minute we see somebody so afflicted. In fact, it is uncomfortable and weird for most downers to use the extensors in their face at all, with the unfamiliar nerve impulses causing the brain a bit of confusion and disorientation when they arrive there. "Dysphoria", it say.

Sexual activity requires a sort of embarrassing extension on the part of the female. While the male will rhythmically flex and extend through an act, the result of which is, possibly, a yoking.

Opening of the eyes, surely, is an act of extension. Closure is flexion. To look up is extension, while looking down must be flexion.

Opening of the mouth is clearly a bit of extension, while few would argue that mouth closure is anything but flexion. Sticking out the tongue is extension, while biting is more like flexion.

You've heard of the gluteus maximus. Uh, it's a flexor (remember, the human had its legs flipped around with the extensors facing front, "turning out" this plump flexor towards the dorsal).

But consider the plight of the gluteus medius muscle, situated as the "side" of the waist, and connecting to the large mass of bone on the femur, the greater trochanter ("the hip"). This muscle must hold the pelvis in place while the opposite leg is lifted from the ground to make stride. In women, the girl pelvis is 1/3 wider, which aids in the wiggle of the walk. With age and time we lose this, shortening the stride and cheapening the gait.

You lose good medius performance along about the time you stop being 'hip". Or is it the converse which is true?

I notice in old people that their pupils are always real small, making it nearly impossible to look in there with an ophthalmoscope. This must mean that the round muscle that contracts and closes the aperture is the flexor, while the long spindly muscles that form the "spokes of the wheel" of the iris are, you guessed it, the extensors. Remember, iridologists look to the splotches in your iris' extensors as if it were a microcosmic exhibit, like the palm. What are they looking at? What are they seeing, by knowing what to look for, in this critical ancient extensor?

Regarding the palm. The palm is a classic ventral structure. Thick, sensitive, bloody, and full of flexion. Don't forget, the hand ends at the elbow. The "back" of the hand and arm show you dorsalness, and extension.

Waving is extension. We reach to shake hands with extension, then consummate the gesture with a little controlled flexion (usually), then "let go" with extension. And then there's the old man who waves to you, with most of the joints of the fingers in flexion. The relative amount of arm waving is an example of how much extension this person wants to (or is able to) put into this greeting. A wave "Hello" is the most obvious everyday illustration of the plight of extension versus flexion.

A punch with the clenched fist is an act of flexion. The backhanded slap is an act of extension.

Internal Flexion

Inside the body we need the work of musculature to do things like move foodstuff through the gut and move blood and maintain blood pressure. This is done with a form of muscular tissue that differs from our skeletal counterpart in microscopic appearance. The fibers generally encircle a roundish tube, as "sphincter" musculature. With coordinated contraction they move a liquid or semisolid that is in their field of squeeze, and this results in movement of this fluid through this tube in one direction.

In the circulatory system, and the tubes we call "vessels" (or vasculature) clearly the action is that of continued flexion. At the center of it is the heart, the poster boy of flexion. The vessels exhibit tone, maintaining a blood pressure, and God forbid any of these lose their flexion because if they do you've had it.

In the gut, the tube does nothing but flex and relax. Like the heart, there is flex, relax, and fill. It's as if the tubular system mandates filling during relaxation, all this taking the place of extension.

At the end of the gut tube is a famous sphincter that, most of the time fortunately, flexes like a champ. But it has a unique action: when pressed upon, it relaxes. It's the only one with this quality. We always knew it was a special flexor.

Having gone from concrete to sublime to ridiculous, a point is there made: among the important dichotomies of the local universe is the concept of physical, bodily opposing forces known as flexion and extension. We can use this paradigm to ask basic scientific questions like "what good can come from tinkering with the forces that drive this opposing system in and out of balance?"

As we, fractals, do our thing, we get there through the separate and individual plights of our many component fractals, and famously among these are things muscular. The better we are able to keep them in some balance, the prettier we will "do". Recognizing, first, that there is a dichotomous relationship between them, and assessing their individual needs, requires an insight few have, and an intervention few are interested in, or capable of, affecting. C'est, folks, is la vie.

Section 4:
Essays, On Humans

"The Spoils of Capitalism"

Look around, and be amazed. So much, so many. The thigamajigs and whatzits. The "stuff". To be so "free", to have and to hold, and to want. Be a glutton, and don't fog up. It's all good.

Careful. You may find yourself to be so dazzled as to miss the marvel. It can all be so distracting as to make a human forget certain whys and wheres. A dilution and perhaps dissipation of wonder, if not wander.

Alas, the spoils of capitalism. The plight of American humanism is the unique legacy of a world's history of life itself, no less, and the great collage and mosaic of the grand diversity of the race. Its "system" of interaction has been based on the one essential of marketeering: obtain currency.

By hook, by crook, by luck, or even by work, to play this game, it takes bucks, and that's all there is to it. In our flavor of economy we call it "the dollar", a note that represents a certain amount of gold (Gasp!), and, well, even though it's a little green on the back, it, uh, resists the oxide pretty well. Just like gold.

In WWII, American productiveness, and productivity, based in capitalism and free markets, turned the tide in the global conflict between subjugation and dictatorship on the one hand, and the power of a cohesiveness of the free on the other. Once the gargantuan "machine" of the US aimed at whatever it decided to aim at, it consistently ensured extinction of expanses of bad conceptualism. Hence, the freedom of freedom itself to spread about the globe has been, certainly, the greatest spoil of its capitalism.

And yet, the results are so mixed. Aside from reduction in morbidities and a fantastic increase in comforts for the creature, there clearly seems to be a toxic component to the system the "older" it gets. The emergence of problems that are unforeseen, of fractals gone unchecked, and butterfly effects encouraged so unwittingly. We add them to the spoils of capitalism, a euphemism and homonym of a close relative of the word, "spoilage". Rotten, gross.

So just try not to mind leaky nuke drums, and broken supertankers. And don't weep at clear-cut forests. And extinctified species. And fished out oceans, and depleted farmlands. Don't choke on the blanded diet you eat, or gag from the dirtying of your liver by fancy chemicals of capitalism. And really, don't be a-scared of the mail, there's just a really small chance it can kill you.

And when disaster strikes, we'll hope to not find out it was the result of some cost-cutting maneuver, or unpaid bill. Or a help too cheap, or a wrong corner cut in an effort to maximize profit for some needy pocket.

Mass communication has ridden the train of modern capitalism to the chronic ache in the psyche of the human soul: massive communicationism. To tune in, or read up, is to be bombarded with all the negative emotions of a multi-faceted survival guiltism, and we forget that if there wasn't no wreck, nobody'd be rubberneckin'. But heck, nobody can resist the ventilation of these natural, survival based, guilty feelings. So the jamming of it down your throat is just a little spoil of that there capitalism, and they know you'll swallow hard. Sells soaps and the like, of course.

We see the negative impact on the thing by lawyerism, which one could argue created capitalism to begin with. We note the drag of the government it hath sustained, and the ironic threat to the vitality of the system by these same professions. Invention, begetting necessity, begets lawyerism. We feel the tug of the madnesses of our reality, the drain on our energies, the grinding of the wheels creaking out the din of the spoiling capitalism.

We rue the continuous invasion of our privacy by the tentacles of the imperialistic machine, in the mailbox, the e-mail inbox, and that little ring box that creases the silence to say, Hello, so-and-so, this is so-and-so calling, and I want some of your money. To enjoy our pastimes, we get splished by so much advertisingism that, well, it hurts. The glare of logos. Parks named ".com". Bowl games, with a first name, effectively assimilating them into "the fam", the sell-out sellouts. And a steady diet of "0% APR, dammit!" Yuk, I've been targetgrouped.

Lotteries? Gaming? The track? The dark side of a money-grub-based society reaches into the pockets of vulnerable people, who are "just havin' a little fun" throwing a fistful of dollars into the glitz of this subliminating, vacuous negative vibration. Heck, for only six or so percent of them will a major psycho-social illness result. For the rest, a little, subtle stumble in the boots of the day tripper. A small ache in the legs wrought by the wear and tear and overuse, on the capitalism exertional.

Golf, they say, is a good walk spoiled. Capitalism, for many, is a good spoil spoiled. We hope for a leadership that is immune to the almost irresistible lure of the cookie jar, and instead see a government bloated, sluggish, and inefficient. We watch global politics where the weight of countries has and always will be measured in their monetaries. We see them fight a war whose perpetrators are trying to get them to "not spend", whether they know it or not. The casualty is the casuality of commerce. An economy in mourning.

Blinded by the blast of modern capitalism, the helpless and hopeless everyman ponders his future like the old sot on the riverbank. The canister that had once held such promise, and so many answers, now spent. Staring on into its empty recesses, he recognizes that needy face in the reflection. He discards it, knowing full well that getting a new one is going to be an adventure.

"The Dummie Syndrome"

And then there is "The Dummie Syndrome". This is the condition that exists when an attempt to do one thing has resulted in the opposite outcome of what was intended. From simple deeds to complex ones, and from personal endeavors to group efforts, often the result is parody.

It isn't hard to do, accomplishing the opposite of your goal. If you stay on a bad journey long enough, the destination could really be off some distance. Why didn't you turn back?

Usually, it's because of some form of self-absorption. This causes a consistent missing of clues and omens, of results real and or imagined. Before long, you suffer the syndrome.

If you're laughing too loud, then probably you're crying inside. People who talk a lot basically have very little to say, while quotable folks are typically parsimonious. And those who want much tend to enjoy little. People who like being around people are the people nobody likes being around. So, do dumb people have more fun? Well, for sure, dummies are no fun, and have no fun.

Excessive primping and otherwise preparing are for those who look at themselves as ugly; you can bet that, inside, they are. They don't know that beauty comes but from the twinkle in the eye.

If you're living out your life vicariously through your kids, everyone including them will learn to hate you for it. Ultimately the family flies apart, instead of hanging together.

Live a lie and the truth will show.

The folks who pray the loudest mean it the least. And if you revel ever so slightly in the misfortune of others, misfortune itself will find you.

If you're first to know the wrong stuff, then you'll be the last to know the right stuff. Poke fun at yourself, and think of yourself as a dummie, and you probably aren't. So people laugh. Poke fun at others, and be assured that the fool is the one that you don't see.

Control "freaks" end up controlling only things that aren't worth controlling. Anybody who says "I ain't gonna lie to you…" is fitting to tell you a half-truth. Rest all with this assurance: hate will rot the vessel that holds it.

If you're a hack, the job itself will defeat you.

You are what you eat; just look at yourself.

So there, just a few simple examples of the syndrome. Perhaps you could name a few, or recall times when things not only went bad, but went all the way bad. Was there something wrong to begin with? And just how powerful is the butterfly?

"The New Enemy: Society Itself"

History doesn't really "repeat" itself, but we note a common theme in so many things, presumably because the same forces at work in one epoch or eon are and will continue to be at work on all things now and hence. We recognize this "rise and fall" of civilizations past. We wonder now about our own.

Perhaps mankind's greatest leap was his ability to integrate as a society to provide for the common good. And provide he has. Having fended off so many threats to his existence, he now faces the reality that he himself poses the greatest danger to all he has become.

While not attempting to downplay the vast complexities involved in the unfolding of events, it is clear that a certain concern or fear can settle into even normal folk regarding the dangers of a human-eat-human existence. With this new set of boogie-men breathing down the necks of the common Joe, more and more of the latter are patterning a behavior of avoidance, inhibition, and selfism to "battle back" at these enemies within.

A look at the culprits is in order.

WANTED: THE LAWMAN Historians will one day relate the profoundly negative impact of the corruption of the legal system on the plight of Americans of The Modern Age. With the help of their ranks in the political "arena", the parasitic infestation of tort has wreaked havoc on the productive sectors of society. Among the drain on their energies is the gut feeling of being "done wrong" by "the system". The fear of the "nobody-everymen" is expressed in righteous indignation by the exasperated good guys. Customers and clients, now "the enemy".

WANTED: THE POLITICIAN There's a saying, "People build, and governments destroy." While most individuals who fancy themselves public servants are decent people, for mis-guided alpha males it's a control thing and a narcissism thing. They lie the best, and so we put them at the top of society's alter ego, its government. The accumulation of such insightless fools has resulted in a gigantic mass of rip-off, squander, and counterproductiveness. Ironically, their detrimental interventions into the plight of society guarantee exactly what we love least, a big huge government. Against pain of prison, they suck life and energy out of his pockets, working him long and extra. Naturally, he sees this federalness as a part of society that is a drag on him, an enemy to him.

WANTED: THE MONEYMAN There are two kinds of people in the world, the kind who collect interest, and the kind who pay interest. The modern American moneyman, with the skillful push of the pen, and the paper, can scratch out millions from thin air. With the slip of a buck from here and there and everywhere, suddenly he's in the "middle" of millions. It is the de-facto dependency of a large percentage of productive dollars, off the top, that keep at bay a group of folks who produce exactly nothing. CEO's, paid in mere millions, reap the benefits of cutting costs and services to and for the working souls who pay the premiums. Such behavior by the megaloes is appropriated by the two previously-mentioned takers, the God-forsaken and whale-in-the-way-of-progress, the middlemen from hell. The pull of the void on decent, productive people is measured not just in bucks for nothing, but in the ache of inconvenience, and in the perception of this element of society as an enemy.

WANTED: MEDIA MONGERS It should fascinate you how amazingly similar the people on the local news look no matter what city you're in. What kind of people are they? They report, mostly, on bad things, and serve as a streaming source of negative energy into the lives of whomever's watching, begging the second question, which is, who's watching this stuff in the first place?

But a lot of people like to "stay informed", so they get with their info anything resembling "horro-drama" that may unfold in any corner of the universe, until eventually you can't help but feel a little scared of what may be lurking. Weirdos, real or imagined, must really walk among us, and even if our relative paranoia edges up ever so slightly because of this spooking, then of course those who "sell" us this "news" are, in this way, contributing to the perception of the average person that yes, the society is a dangerous thing, an enemy.

For anyone whose perception is their reality, they'll hallucinate a few more enemies than the rest, and alter their behavior to dodge them. In so doing, he will pose the biggest problem of this burden, where he loses his savvy enough to become, himself, his own, worst, enemy.

"Don't Feed The Humans"

Critical to the understanding of these units, humans, is to have a feel for the metabolic machinery of their body, the chemical reactions of life. When you look upon such existence as what it is, a slow burning fire that eventually burns out, you'll start conjuring up concepts on just how it ought to be stoked.

The main ingredient in any fire is that most hungry of elements we introduce through the lungs, oxygen. The better you can breathe it, the better you can circulate and deliver it, the cleaner the cooker will be, and the longer your tissues will hold up. Sluggish supply of this gas, due in general to poor cardiovascular conditioning, is the main enemy to the goal of clean, work-a-day metabolism. And the main enemy to the heart and its cast is being overweight, a condition that comes, simply, from overfeeding of the human.

Byproducts of the body's energy chemistry, rascal molecules referred to as "free-radicals", are unstable collections of atoms that "pluck" what they need from local proteins. Nutrients like vitamins can serve to "gobble" free radicals, thereby protecting the body's vital structures. Aging phenomena from skin wrinkling to dementia to hearing loss and macular degeneration (old-people blindness) are all "wear-outs" from decades of free-radical attack.

The best way to limit the volume and additive number of free radicals is to burn a quieter fire. Whether you're counting fat grams or carb violations, you should never forget that anything you stuff down the top of your humanity is another inching along of the finite burn and cook. So if the scale is tipping ever so gradually upward, your jaunt and gallop down this road is a more significant lope. Be not surprised, then, that the darned thing burned hotter cause you fueled it so.

The manner in which cigarette smoking accelerates the aging process uncovers life for what it is, this long burn. By adding a vicious free radical, carbon monoxide, to the band of assassins, the tick of the clock is ratcheted several clicks. The gassing of the lungs also diminishes the entrance of oxygen into the bloodstream, and of course oxygen is the ultimate free radical scavenger. Furthermore, nicotine constricts the body's microcirculation, further toxifying and concentrating sluggish metabolism where we want it least, at the cellular level itself. So, from bad skin and teeth to a rusted set of pipes, an acceleration of aging, indeed, on display for you, by the human feeding him and herself this awful "fix" of this wholly silly "fit". If they look older than their age, it's because they are.

To drive out west, in the vast expanses of America, is to witness the marvels of the taming of the land. Far as the eye can see, cattle and carbs. It's almost impossible to fathom the expanse of wheat and corn raised by our farmers and their machines. And just because a lot of the population is dying of carb poisoning doesn't at all reduce their accomplishments. All anybody ever wanted was to feed the humans.

Diabetes mellitus, the latest "epidemic" enabled by the ridiculosity of our affluence, results from overuse and eventual depletion of the body's insulin mechanism. Insulin, the only hormone that stores fat and sugar for the body, simply isn't used to storing and storing and storing energy. As the tissues run at full capacity for so long, they become "resistant" to the efforts of insulin to jam them with more. So the pancreas secretes more insulin to make up for this, and by the fourth or fifth decade it just can't do it any more.

When insulin levels fall, the condition of diabetes is marked by skyrocketing of blood levels of fat and sugar. Over time, this is a metabolism that causes, in no particular order, heart attacks and strokes, blindness, a diffuse neuropathy, impairment of white blood cell performance (infections), amputations, and kidney failure to the point of dialysis. Diabetics burn so hot you can almost feel it and see it.

The condition has only one real treatment: not-feeding of the human.

Well conditioned, the human is the king of the jungle. His crafty mind, his nimble fingers, and an impressive size and power allow him to shimmer the hooves of all the pretenders. But hah! Seen him lately? He fights the neglect of his muscular system all day every day, and knows full well that his thirst and its quenches, and his hunger and its satiates, can create no quirks his muscularity can't handle for him. Still, the sofa beckons, and now look at him. Just look at him.

Girls really need to watch it. Somehow in them eating is different. Hunger is more tied into emotions, and becomes a sort of sport or preoccupation for them. When they're sad and unsure, they eat. They may even complain that they're "starving". Or if they were wrong-nurtured, they might end up either making themselves puke, or just not-feeding altogether, as an identifiable behavioral disorder. So a simple question might be, why is foodstuffing even an issue?

Men and women, so different, also have this opposite-ness: men store their fat centrally, in their belly and organs, while women store it peripherally, in their skin. For both, overweightness and obesity result in parody figures, with the male's physical deftness and "shape" impaired by the over-represented torso, and the gal looking at beauty-loss through chubby cheeks, rolls in the midriff, and girthy extremities.

The treatment for both is the same. They share a same philosophy, of denial that life is harder than they're pushing, and an inhibition that makes them fill their stomachs rather than their gym bags.

New epidemic: reflux. Acidic stomach juices injure the bottom of the esophagus if they "reflux" backward. Ideally, things in the gut tube go the other direction. But something has to promote the digestive system to move things along. Nothing does it like exercise. Enough said.

To treat this most prevalent condition, a billion dollar pharmaceutical: drugs that make the stomach stop secreting acid. They work. The stomach juices still reflux backward into the swallowing tube, but without the acid in them, the injury is far less. It will figure, then, that eventually humans will be born with no acid-secreting cells in their stomach. And maybe a remote in their palm.

Want to play good at a sport? Don't eat for most of a shift before gametime. A little starvation is great for the mind. Great. The reflexes are quickened. The desire is more natural. Get your bone, dog, and they'll feed ya later.

Maybe, if you could hop on a Starwisp, and travel the galaxy, you would come upon a great galactic zoo, one that featured a vast assortment of goons and kooks and freaks from deep into the parsecs. They, too, have implored their physics, and explored their math, and got loose. How they ended up there is, well, it's a matter of history and crowded circumstance you might say. Would there be humans?

Look, there. Hoppin' and bobbin', and poppin' off at their social. Painting their faces, playing tunes, and hanging. Natural environs. Healthier, niftier, and shiftier. And a whole lot smellier, that's for sure.

And there, a sign. It says, "Please, Don't Feed The Humans".

"Life's Hard, Not Easy"

Oh, the games we play. The simple things we don't say. The simple dreams, we let them slip away.

The field of dreams is mined with mire and muck, and plenty of turtles to trip you up. You can expect no easy nothing. If you cut corners, then you might just as well cut your big toe off. A stitch in time saves you a minute or two max. Skip a grade and you're in a big hurry to no-where.

Paying attention is only somewhat helpful. Blinders can be of great benefit, though you may not ascertain this initially. The recommendation: Keep your head down and stay on it; there is no such thing as luck.

The planet "Earth" offers a continual demonstration that life is hard and not easy, and yet casual observers can miss the calamity unfolding before them. A tree falls in a forest, yet only its most keen observers will hear the noise, whilst the deaf argue whether there was sound, what with them not there to hear it.

And not just trees, for gravity brings it all down, from water droplets to mighty mountains, and there is noise. It is the crash and bang of noteasyhood.

Meanwhile the snow appears to be going up, yet everything's turning white. Let's ask the experts.

Prey, who generally keep to themselves, know to expect predation. Innocently spectacular, plant life can expect the same. At times, hard times for all, for sure.

Eventually the Big Bell rings. It is the clang of hard reality that overshadows the hardness of reality.

The human body is beautiful. After a brief cultivation (which is hard, not easy) it blossoms into what looks like perfection. Interestingly, the harder it is worked the more beautiful it becomes. And while the bloom fades ultimately, it will do so with much more grace if this type of attention is consistently paid to it. For any number of wrong reasons, this turns out to be too hard for most, and the beauty dies prematurely, a tragic thing. These types of reasons exemplify violation of the rules of this game previously stated.

There is such value in toil. Work equals force times distance. Power is work per unit time. Since we know that time and distance are relative, possibly, we could postulate a few things about work-force-power. They are the hammer pounding in the nails of life's hardness.

And we don't think of ourselves as "lucky" if we, say, discover something by accident, because a lot of great things were discovered that way. The hard way.

People say they wish they could play a musical instrument, and musicians know why they don't. It's all those years of practice it takes to be able to play it good. And all the nerve it takes to pour your heart out. And you suck for so long. That's hard.

For stars, the fall can be far. All the hard work it takes to be at the top of anything can be easily forgotten (apparently) when it comes time to realize that staying on top and wanting to stay on top are both hard and not easy. Once back to the bottom, the knock on the head reminds them of that which they once knew so well.

It's easy for a zillionaire to concoct a squad of can't-miss players, and to have great expectations, yet it's the magical mix of wise veterans and talented, enthusiastic youngsters that provides the synergistic "chemistry" needed to win. We can say that, in this way, alchemy went out of style along about the time humans first discovered that life is hard, and not easy.

Throughout our society we have favored easy solutions and quick fixes to problems we don't even understand to begin with, and in most cases have managed to make such matters worse. We know all this. We know when sources of negative energy cut us to the quick, yet our questions are few, our answers somehow many.

Again the tree falls, and not a sound. Now the sharks are gone. Gasp! No lions? No leopards? The great monkey escape of the twentieth century begs a question: how much more of us will this blue planet put up with? And what does all this have to do with life being hard and not easy?

More mansions and less forest. More freedom and dirtier spirits. Different times, same species, still plundering and blundering all the while, as if conserved like mass and energy, the same old story of life, that it is hard and not easy.

About the Author

Daniel A. Shields, MD is a trained Family Physician who has worked in the Emergency Department full-time at a very busy modern community hospital in America's midwest. At the time of this writing he was in his 17th year doing what is surely the hardest job in the city. The congeniality and normalness reflected in this literary art is the same social skill that has allowed him special access to the struggles of his fellow man. "Hence this volume."